6 THEA Math Practice Tests

Extra Practice to Achieve a Crack Score

By

Elise Baniam & Michael Smith

THEA Practice Tests

6 THEA Math Practice Tests
Published in the United State of America By
The Math Notion
Email: info@mathnotion.com
Web: www.mathnotion.com

Copyright © 2021 by the Math Notion. All rights reserved. No part of this publication may be reproduced, stored in a retrieval system, or transmitted in any form or by any means, electronic, mechanical, photocopying, recording, scanning, or otherwise, except as permitted under Section 107 or 108 of the 1976 United States Copyright Ac, without permission of the author.
All inquiries should be addressed to the Math Notion.

ISBN: 978-1-63620-186-3

About the Author

Elise Baniam has been a math instructor for over a decade now. She graduated in Mathematics. Since 2006, Elise has devoted his time to both teaching and developing exceptional math learning materials. As a math instructor and test prep expert, Elise has worked with thousands of students. She has used the feedback of her students to develop a unique study program that can be used by students to drastically improve their math score fast and effectively.

- **SAT Math Workbook**
- **ACT Math Workbook**
- **GRE Math Workbook**
- **HSPT Math Workbook**
- **Common Core Math Workbook**
- **many Math Education Workbooks**
- **and some Mathematics books ...**

As an experienced Math teacher, Mrs. Baniam employs a variety of formats to help students achieve their goals: she teaches students in large groups, and she provides training materials and textbooks through her website and through Amazon.

You can contact Elise via email at:
Elise@mathnotion.com

THEA Practice Tests

6 Practice Tests to Help Achieve an Excellent THEA Math Score!

Practice makes perfect, and the best way to exercise your THEA test-taking skills is with simulated tests. Our experts selected these targeted questions to help you study more realistically and use your review time wisely to reach your best score. These math questions are the same as the ones you will find on the THEA test, so you will know what to expect and avoid surprises on test day.

6 THEA Math Practice Tests provide six full-length opportunities to evaluate whether you have the skills to ace the test's higher-level math questions.

This book emphasizes that any difficult math question focuses on building a solid understanding of basic mathematical concepts. Inside the practice math book, you will find realistic THEA math questions and detailed explanations to help you master your math sections of the THEA. You will discover everything you need to ace the test, including:

- Aligned to the latest THEA test.
- **Fully explained answers to all questions.**
- Practice questions that help you increase speed and accuracy.
- Learn fundamental approaches for achieving content mastery.
- Diagnose and learn from your mistakes with in-depth answer explanations.

With the THEA math prep, the lots of test takers who would like an intensive drill with multiple math questions, get a quick but full review of everything on their exam. Anyone planning to take the THEA exam should take advantage of math practice tests. Purchase it today to receive access to THEA math practice questions.

WWW.MATHNOTION.COM

… So Much More Online!

✓ FREE Math Lessons

✓ More Math Learning Books!

✓ Mathematics Worksheets

✓ Online Math Tutors

For a PDF Version of This Book

Please Visit www.mathnotion.com

THEA Practice Tests

Contents

THEA Test Review 9
    THEA Test Mathematics Formula Sheet... 11
    THEA Practice Test 1 13
    THEA Practice Test 2 31
    THEA Practice Test 3 49
    THEA Practice Test 4 67
    THEA Practice Test 5 85
    THEA Practice Test 6 103

Answers and Explanations 121
    Answer Key... 123
    THEA Practice Test 1 127
    THEA Practice Test 2 135
    THEA Practice Test 3 142
    THEA Practice Test 4 150
    THEA Practice Test 5 157
    THEA Practice Test 6 165

THEA Test Review

THEA Test Mathematics Formula Sheet

Area of a:	
Parallelogram	$A = bh$
Trapezoid	$A = \dfrac{1}{2}h(b_1 + b_2)$

Surface Area and Volume of a:		
Rectangular/Right Prism	$SA = ph + 2B$	$V = Bh$
Cylinder	$SA = 2\pi rh + 2\pi r^2$	$V = \pi r^2 h$
Pyramid	$SA = \dfrac{1}{2}ps + B$	$V = \dfrac{1}{3}Bh$
Cone	$SA = \pi rs + \pi r^2$	$V = \dfrac{1}{3}\pi r^2 h$
Sphere	$SA = 4\pi r^2$	$V = \dfrac{4}{3}\pi r^3$

(p = perimeter of base B; $\pi = 3.14$)

Algebra	
Slope of a line	$m = \dfrac{y_2 - y_1}{x_2 - x_1}$
Slope-intercept form of the equation of a line	$y = mx + b$
Point-slope form of the Equation of a line	$y - y_1 = m(x - x_1)$
Standard form of a Quadratic equation	$y = ax^2 + bx + c$
Quadratic formula	$x = \dfrac{-b \pm \sqrt{b^2 - 4ac}}{2a}$
Pythagorean theorem	$a^2 + b^2 = c^2$
Simple interest	$I = prt$ (I = interest, p = principal, r = rate, t = time)

THEA Practice Test 1

Mathematics

Total Number of Questions: 50 Questions

Total time: 240 Minutes (All three sections)

You may use a non-programmable calculator for this test.

Administered *Month Year*

THEA Practice Tests

1) How many odd integers are between $\frac{-46}{5}$ and $\frac{57}{7}$?

   A. 6

   B. 7

   C. 9

   D. 10

2) Which expression correctly represents the distance between the two points shown on the number line?

   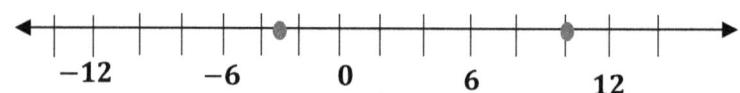

   A. $-3 - 10$

   B. $|-3 + 10|$

   C. $-3 + 10$

   D. $|3 + 10|$

3) In the triangle ABC, if angle B and angle C both equal 60°, then what is the length of side BC?

   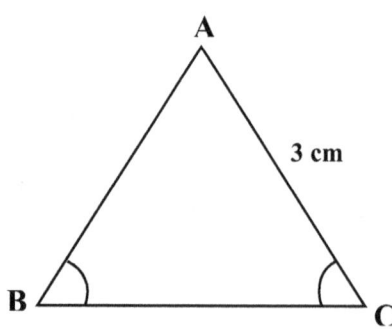

   A. 6 cm

   B. 3 cm

   C. 15 cm

   D. 9 cm

THEA Practice Tests

4) Which of the following represents the sum of the factors of 12?

   A. 28

   B. 30

   C. 18

   D. 26

5) Arrange the following fractions in order from least to greatest.

$$\frac{5}{9}, \frac{2}{5}, \frac{1}{8}, \frac{25}{28}, \frac{12}{17}$$

   A. $\frac{1}{8}, \frac{2}{5}, \frac{5}{9}, \frac{12}{17}, \frac{25}{28}$

   B. $\frac{2}{5}, \frac{5}{9}, \frac{1}{8}, \frac{12}{17}, \frac{25}{28}$

   C. $\frac{25}{28}, \frac{12}{17}, \frac{2}{5}, \frac{5}{9}, \frac{1}{8}$

   D. $\frac{12}{17}, \frac{25}{28}, \frac{1}{8}, \frac{5}{9}, \frac{2}{5}$

6) Elena earns $6.5 an hour and worked 36 hours. Her brother earns $9.75 an hour. How many hours would her brother need to work to equal Elena's earnings over 40 hours?

   A. 19.5

   B. 24

   C. 30

   D. 32

WWW.MathNotion.com

7) In a library, 20% of the books are fiction and the rest are non-fiction. Given that there are 1,500 more non-fiction books than fiction books, what is the total number of books in the library?

   A. 3,000

   B. 4,500

   C. 2,500

   D. 1,600

8) A map has the scale of 6 cm to 1 km. What is the actual area of a lake on ground which is represented as an area of 90 cm^2 on the map?

   A. 2.4 km^2

   B. 2.5 km^2

   C. 18 cm^2

   D. 2.8 cm^2

9) $15.15 \div 0.3 =$

   A. 5.05

   B. 50.50

   C. 50.05

   D. 5.005

10) What is the circumference of a circle with a radius of 9 inches?

   A. 81π

   B. 18π

   C. 9π

   D. 36π

11) Alfred needs to calculate his monthly water bill. His family used 31,500 gallons at a rate of $0.86 per hundred gallons. Also, there is a monthly fee of $5.30 on each period. What is his total bill?

   A. $3,325.20

   B. $367.30

   C. $276.2

   D. $27.62

12) Simplify $\frac{(3x^4 - 6x^3)}{(x^3 - 2x^2)} = ?$

   A. $3x$

   B. $x - 2$

   C. $6x$

   D. $2x(x - 1)$

THEA Practice Tests

13) Which of the following expressions is undefined in the set of real numbers?

   A. $\sqrt[2]{121}$

   B. $\sqrt[3]{-64}$

   C. $\sqrt{-25}$

   D. $\sqrt[4]{81}$

14) If $f(x) = 6x^2$, and $4f(3a) = 864$ then what could be the value of a?

   A. -2

   B. -4

   C. 4

   D. 2

15) Speed of a train is 140 miles per hour for 3 hours and 25 minutes. How many miles did the train travel?

   A. 455

   B. 285

   C. 355

   D. 265

16) Solve $5x - 8 \leq 12x + 6$

   A. $x \leq -2$

   B. $x \geq -2$

   C. $x \leq 2$

   D. $x \geq 2$

17) What is the value of 2^8?

   A. $(2+2)^8$

   B. 4^4

   C. $2(2^4)$

   D. $2^4 + 2^4$

18) Shane Williams puts $4,200 into a saving bank account that pays simple interest of 3.8%. How much interest will she earn after 5 years?

   A. $2,780

   B. $ 1,198

   C. $798

   D. $189

19) Three angles join to form a straight angle. One angle measure 61°. Other angle measures 42°. What is the measure of third angle?

   A. 17°

   B. 27°

   C. 103°

   D. 77°

THEA Practice Tests

20) Evaluate $\dfrac{15x^7y^4z^{-4}}{6x^3y^8z^2}$.

A. $\dfrac{5x^4y^4}{2z^6}$

B. $\dfrac{5x^3z^2}{2y^2}$

C. $\dfrac{5x^4}{2y^4z^6}$

D. $\dfrac{5y^4}{2x^4z^6}$

21) Find the equation for line passing through $(1, -2)$ and $(4, 3)$.

A. $-5x - 3y = 11$

B. $3y - 5x = -29$

C. $-5x + 3y = 19$

D. $3y + 5x = -21$

22) If $x = -3$ and $y = 5$, calculate the value of $\dfrac{x^2 + 12}{y - 2}$.

A. $\dfrac{1}{7}$

B. -7

C. 7

D. 5

THEA Practice Tests

23) The table shows the parking rates for the outside terminal area at an airport.

Ethan parked at the lot for $4\frac{1}{2}$ hours. How much did he owe?

A. $6.95

B. $7.55

C. $7.95

D. $6.55

3 hours	$4.2
Each 30 minutes after 3 hours	$1.25
24-hours Discount rate	$50

24) What is the value of x in term of c and d $\frac{c-d}{dx} = \frac{3}{5}$? (c and d>0)

A. $\frac{5}{3}(\frac{c}{d} - 1)$

B. $\frac{5}{3}(1 - \frac{c}{d})$

C. $\frac{3}{5}(\frac{c}{d} - 1)$

D. $\frac{3}{5}(1 - \frac{c}{d})$

25) Factor the equation $4x^4 - x^2$.

A. $2x^2(2x + 1)$

B. $x^2(2x + 1)(x - 2)$

C. $x^2(2x - 1)(x + 2)$

D. $x^2(2x - 1)(2x + 1)$

26) Which of the following are the solutions to the equation $x^2 - 16x + 63 = 0$?

   A. $-9, 3$

   B. $9, 7$

   C. $-3, -9$

   D. $9, -7$

27) Which of the following equations best represents the line in the graph below?

   A. $y = \frac{1}{4}x + 1$

   B. $y = x + 4$

   C. $y = \frac{1}{4}x - 1$

   D. $y = 4x + 1$

   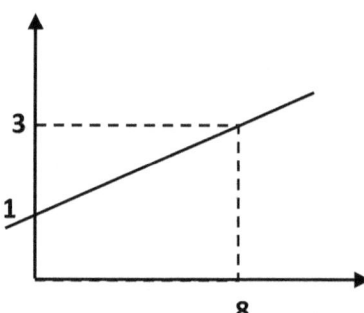

28) If $-4x + 5y = -4$ and $3x - 2y = 3$, what is the value of x?

   A. 3

   B. 0

   C. -3

   D. 1

29) lengths of Two sides of a triangle are 7 and 4. Which of the following could Not be the measure of third side?

   A. 5

   B. 2

   C. 12

   D. 14

30) The following data set is given: 121, 149, 136, 174, 167, 129.

   Adding which number to the set will increase its mean?

   A. 142

   B. 145

   C. 141

   D. 151

31) An award for best education improvement is awarded annually to a winning US state, and the winners from 2002 to 2011 are given in the table below. Find the mode of this set of states.

Year	2002	2003	2004	2005	2006	2007	2008	2009	2010	2011
State	New Jersey	Ohio	Oregon	New York	Oregon	California	Ohio	Oregon	Ohio	New Jersey

   A. New York and Oregon

   B. California and New Jersey

   C. Ohio and Oregon

   D. Oregon and California

32) How long is a distance of 9 km if measured on a map with a scale of 1:60,000?

   A. 18

   B. 15

   C. 9

   D. 5

33) What is the area of a square if its side measures $\sqrt{11}$ m?

A. $\sqrt{11}$

B. $2\sqrt{11}$

C. $3\sqrt{11}$

D. 11

34) What is the answer of the following equation $3x^3y^2 (2x^2y^2)^5 =?$

A. $64x^{13}y^{12}$

B. $96x^{13}y^{12}$

C. $96x^{12}y^{13}$

D. $64x^{12}y^{13}$

35) The circle graph below shows the type of pizza that people prefer for lunch. If 180 people were surveyed, how many people preferred Meat lovers?

A. 26

B. 28

C. 27

D. 24

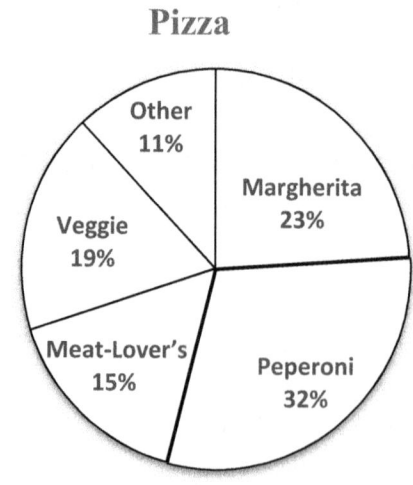

THEA Practice Tests

36) Rosie is *x* years old. She is 6 years older than her twin brothers Milan and Marcel. What is the mean age of the three children?

   A. $x + 4$

   B. $x - 9$

   C. $x + 9$

   D. $x - 4$

37) a is inversely proportional to $(3b - 8)$. If a = 12 and b = 6, express a in terms of b.

   A. $a = 3b - 14$

   B. $a = 120(3b - 8)$

   C. $a = \frac{120}{3b-8}$

   D. $a = \frac{3b-8}{120}$

38) A baseball has a volume of 972π. What is the length of the diameter?

   A. 27

   B. 18

   C. 9

   D. 15

39) 168 is What percent of 120?

   A. 140 %

   B. 40 %

   C. 60 %

   D. 160 %

40) A phone manufacturer makes 15,000 phone a year. The company randomly selects 300 of the phones to sample for inspection. The company discovers that there are 4 faulty phones in the sample. Based on the sample, how many of the 15,000 total phones are likely to be faulty?

A. 400

B. 30

C. 150

D. 200

41) Emma and Mia buy a total of 25 books. Emma bought 3 more books than Mia did. How many books did Emma buy?

A. 13

B. 14

C. 16

D. 7

42) A bag contains 3 white balls, 7 red balls and 4 black balls. A ball is picked from the bag at random. Find the probability of picking a red ball?

A. 1.20

B. 0.20

C. 0.05

D. 0.50

THEA Practice Tests

43) The radius of the following cylinder is 3 inches, and its height are 7 inches.

What is the surface area of the cylinder in square inches? (π=3.14)

A. 188.4

B. 376.8

C. 9.42

D. 37.68

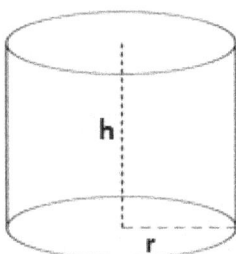

44) The line n has a slope of $\frac{a}{b}$, where c and d are integers. What is the slope of a line that is perpendicular to line n?

A. $\frac{a}{b}$

B. $-\frac{a}{b}$

C. $\frac{b}{a}$

D. $-\frac{b}{a}$

45) In a store, 22% of customers are female. If the total number of customers is 800, then how many male customers are dealing with the store?

A. 286

B. 864

C. 624

D. 426

46) If 72.3 kg is divided into two parts, in a ratio of 7:4, how many kg is the smaller share?

A. 2.96 kg

B. 26.29 kg

C. 2.62 kg

D. 29.26 kg

47) Express as a single fraction in its simplest form: $\frac{3}{(x-1)} - \frac{4}{(2x+3)} = ?$

A. $\frac{2x+13}{(x-1)(2x+3)}$

B. $\frac{2x-13}{(x-1)(3x+2)}$

C. $\frac{2x+13}{2x+3}$

D. $\frac{-2}{2x-13}$

48) Ryan is x years old, and her sister Mitzi is $(7x - 16)$ years old. Given that Mitzi is triple as old as Ryan, what is Mitzi's age?

A. 8

B. 10

C. 12

D. 16

THEA Practice Tests

49) The set of possible values of p is {4, 8, 12}. What is the set of possible values of h if $4h = 3p + 4$?

   A. {4,10,12}

   B. {7,10,12}

   C. {4, 7, 10}

   D. {7,12,14}

50) In the infinitely repeating decimal below, 7 is the second digit in the repeating pattern. What is the 440st digit? $\frac{1}{13} = \overline{0.076923}$

   A. 7

   B. 6

   C. 0

   D. 3

"End of THEA Practice Test 1."

THEA Practice Test 2

Mathematics

Total Number of Questions: 50 Questions

Total time: 240 Minutes (All three sections)

You may use a non-programmable calculator for this test.

Administered Month Year

THEA Practice Tests

1) What is the sum of the smallest prime number and six times the largest negative even integer?

   A. −6

   B. 4

   C. −10

   D. −8

2) Sam's incomes and expenditures for the first season of the last year are given in the table below. In which month were her savings the highest?

   A. February

   B. March

   C. January

   D. All months were the same.

Month	Income	Cost
January	$2,780	$1,002
February	$2,895	$975
March	$2,990	$1,012

3) What is the number a, if the result of adding a to 38 is the same as subtracting $5a$ from 272?

   A. −49

   B. 57

   C. 127

   D. 39

THEA Practice Tests

4) Solve these fractions and reduce to its simplest terms: $7\frac{5}{32} - 4\frac{1}{4} + 3\frac{3}{8} =$

   A. $6\frac{9}{32}$

   B. $-6\frac{5}{32}$

   C. $\frac{5}{8}$

   D. $6\frac{3}{4}$

5) Calculate, $5^2 + 5 + 5^0 = ?$

   A. 5^2

   B. 35

   C. 31

   D. 75

6) Which decimal is equivalent to $\frac{133}{190}$?

   A. 0.107

   B. 1.07

   C. 0.07

   D. 0.7

7) The price of a shirt increased from $40 to $42.40. What is the percentage increase in the price?

   A. 6%

   B. 0.6%

   C. 0.94%

   D. 1.6%

THEA Practice Tests

8) Find the solution set of the following equation: $|3x - 6| = 9$

   A. $\{5, -1\}$

   B. $\{1, -5\}$

   C. $\{1\}$

   D. $\{-5, -1\}$

9) What is the solution to the pair of equations below? $\begin{cases} x - 4y = 6 \\ 3x + y = 5 \end{cases}$

   A. $x = 1$ and $y = -2$

   B. $x = -1$ and $y = 2$

   C. $x = 2$ and $y = -1$

   D. $x = 2$ and $y = -2$

10) The train each 20 minutes passes an average 3 stations. At this rate, how many stations will it pass in four hours.

    A. 24

    B. 36

    C. 18

    D. 72

11) If Sofia buys a shirt marked down 12 percent from its $170 while Natalia buys the same shirt mark down only 10 percent, how much more does Natalia pay for the shirt?

A. $0.34

B. $1.70

C. $17

D. $3.40

12) Find the circumference of the circle in terms of π.

A. 14.5π in.

B. 29π in.

C. 58 π in.

D. 116π in.

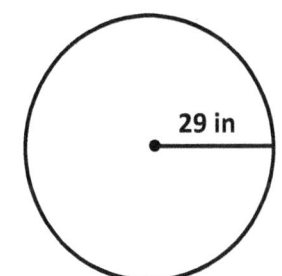

13) Find the volume of rectangular prism below?

A. 980 in^3

B. 1,090 in^3

C. 2,990 in^3

D. 5,880 in^3

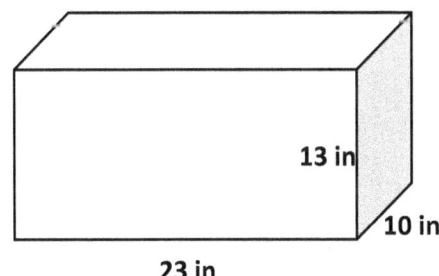

THEA Practice Tests

14) What is $\sqrt[3]{3^{-9}}$ in simplest form?

A. $\dfrac{1}{2,187}$

B. $\dfrac{1}{729}$

C. $\dfrac{1}{9}$

D. $\dfrac{1}{27}$

15) Which statement correctly describes the value of N in the equation below?

$5(9N - 11) = 9(5N - 13)$?

A. N has no correct solutions.

B. N has infinitely many correct solutions.

C. N=1 is one solution.

D. N=0 is one solution.

16) What is the maximum amount of grain, the silo can hold, in cubic feet?

A. $480\pi\ m^3$

B. $148\pi\ m^3$

C. $408\pi\ m^3$

D. $1{,}630\pi\ m^3$

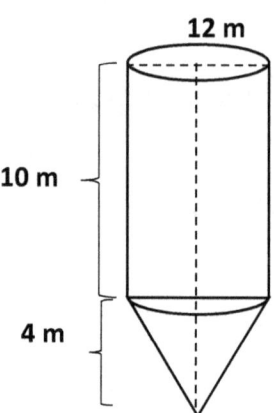

THEA Practice Tests

17) What is 7.59×10^{-3} in standard form?

   A. −75,900

   B. −0.00759

   C. $\dfrac{1}{75,900}$

   D. 0.00759

18) What is the value of x in the triangle?

   A. 44°

   B. 134°

   C. 46°

   D. 131°

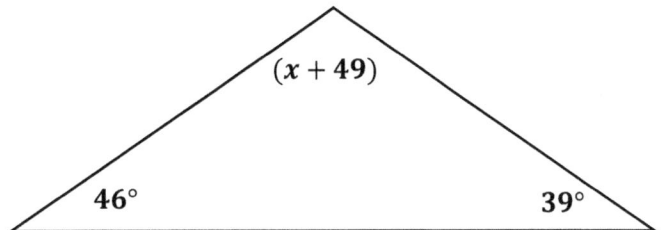

19) Find the length of the unknown side.

   A. 28 ft

   B. 24 ft

   C. 22 ft

   D. 26 ft

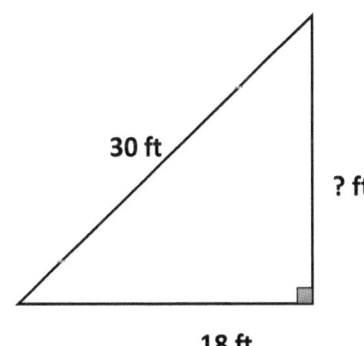

20) A store sells all of its products at a price 15% greater than the price the store paid for the product. How much does the store sell a product if the store paid $160 for it?

A. $248

B. $124

C. $184

D. $24

21) What is the area of shaded region?

A. 68

B. 92

C. 24

D. 106

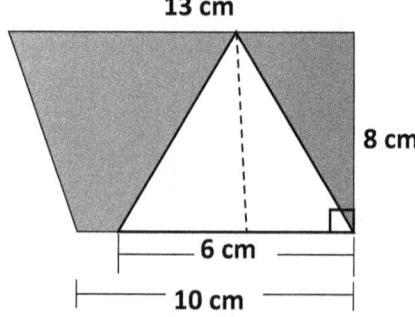

22) if $xy - 9x = 42$ and $y - 9 = 7$, then $x =$?

A. 4

B. 12

C. 7

D. 6

23) Which is the value of x in the equation $\frac{x}{5} = x - 8$?

A. 5

B. 13

C. 10

D. 8

24) What is the value of x, If $-7x + 3y = 8$ and $-6x + 5y = 2$?

A. -8

B. -1

C. -6

D. -2

25) What is the probability of Not spinning at H?

A. $\frac{1}{2}$

B. $\frac{1}{8}$

C. $\frac{1}{4}$

D. $\frac{3}{8}$

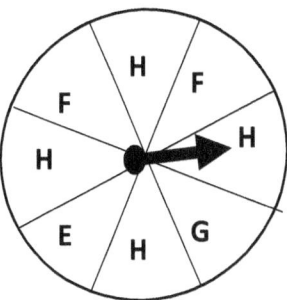

26) Ella bought 42 movies for $4.06 per movie. Which equation shows the BEST estimate of the total cost?

A. 40 × $4 = $160

B. 40 × $5 = $200

C. 42 × $4 = $168

D. 42 × $5 = $210

27) A pick-up truck travels 60 mile on 9 L of gasoline when driven on a smooth road. If the cost of gasoline is $1.20/L, which is the cost of 1,400 mile of highway (smooth)?

   A. $151

   B. $1,255

   C. $125.5

   D. $252

28) A position of subway station and Grace's house shown by a grid. The station is located at $(-3, -5)$, and her house is located at $(1, -2)$. What is the distance between her house and the subway stop?

   A. 4

   B. 5

   C. $5\sqrt{2}$

   D. 3

29) For the following set of numbers find the median.

   230, 98, 389, 156, 215, 188, 401.

   A. 215

   B. 156

   C. 230

   D. 188

THEA Practice Tests

30) What is solution to the equation $\sqrt{5x-6} = 7$?

   A. -6

   B. -12

   C. 8

   D. 11

31) The equation $x = 3y - 9$ has a y-intercept of?

   A. 3

   B. -3

   C. $\frac{1}{3}$

   D. $-\frac{1}{3}$

32) If $2^{3x} = 64$, then $x = ?$

   A. 8

   B. 1

   C. 2

   D. 6

33) Which is the smallest positive integer which is divisible by both 32 and 48.

   A. 12

   B. 48

   C. 96

   D. 120

34) A cube has total surface area of 150 cm². what is the volume of the cube in cm³?

A. 5

B. 125

C. 25

D. 75

35) Which is the value of x^2, if $x^2 + 4x = 32$?

A. 9

B. -16

C. 49

D. 64

36) Each of 8 pitchers can contain up to $\frac{7}{8}$ L of water. If each of the pitcher is at least the one-fourth full, which of the following expressions represents the total amount of water, W, contained on all 8 pitchers?

A. $0.7 < w < 7$

B. $1.25 < w < 7$

C. $0.75 < w < 3.5$

D. $1.75 < w < 7$

37) Ages of the players on a volleyball team is given. Which is the range of their ages? 27, 42, 19, 51, 25, 50, 22, 37, 30, 16, 46, 29

   A. 28

   B. 16

   C. 35

   D. 51

38) Find the area of the circle to the nearest tenth. Use 3.14 for π.

   A. 379.9

   B. 549.7

   C. 68.0

   D. 44

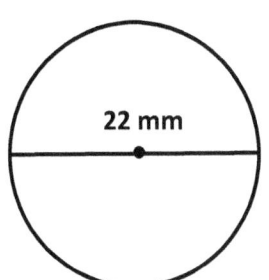

39) What is the simplest form of the expression $\dfrac{3x^2-14x-5}{9(x^2-\frac{1}{9})}$?

   A. $\dfrac{x+5}{3(x-\frac{1}{3})}$

   B. $\dfrac{x-5}{3x-1}$

   C. $\dfrac{x+4}{3x+1}$

   D. $\dfrac{x-2}{x-3}$

THEA Practice Tests

40) What is the value of $\left(\frac{1}{2}\right)^{-5}$?

A. $\frac{1}{32}$

B. -32

C. $-\frac{1}{32}$

D. 32

41) What is the simplest form of the expression $\frac{(6x^{-4}y^2)^2}{252y^{-3}z^0}$, (using positive exponent)?

A. $\frac{y^7 z}{7x^8}$

B. $\frac{y^7}{7x^8}$

C. $\frac{7x^8}{zy^8}$

D. $\frac{x^8}{7y^7}$

42) Some fruit sells for $20 per kilograms. What is the price in cent per gram?

A. 0.002

B. 0.2

C. 0.02

D. 2

THEA Practice Tests

43) The price of water quadruples every 5 years. If the price of water on January 1st, 2012 is $8 per gallon, what is the equation that would be used to calculate the price(P) of water on January 1st, 2007?

A. $4P = 5$

B. $\frac{P}{8} = 4$

C. $4p = 8$

D. $8P = 4$

44) The line $3y + 4 = 18x + 7$ and $6y - 2 = x + 3$ are:

A. The same line

B. Parallel

C. Perpendicular

D. Neither parallel nor perpendicular

45) A rectangular box measures $7\frac{1}{3}$ feet by $6\frac{3}{4}$ feet. It is divided into three equal parts. What is the area of one of those parts?

A. 33

B. 8.5

C. 16.5

D. 49.5

THEA Practice Tests

46) Find the slope of the line.

A. $\frac{1}{2}$

B. $\frac{1}{4}$

C. 4

D. $-\frac{1}{3}$

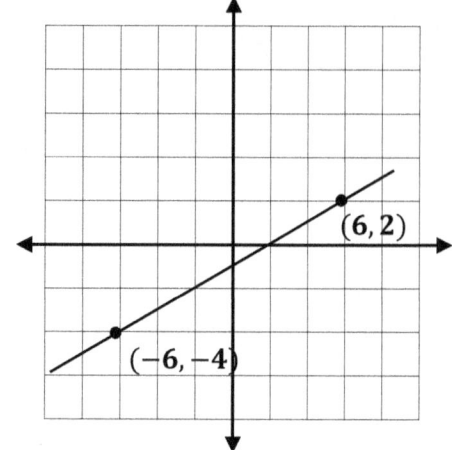

47) Amelia cuts a piece of birthday cake as shown below. What is the volume of the piece of cake?

A. 508 cm^3

B. 408 cm^3

C. 480 cm^3

D. 1,480 cm^3

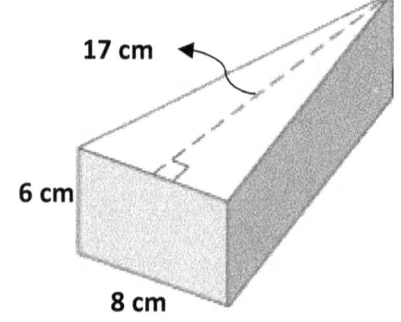

48) Let $f(x) = 3x - 6$. If $f(a) = -15$ and $f(b) = 6$, then what is $f(a + b)$?

A. 6

B. 3

C. −3

D. −6

WWW.MathNotion.com

49) Given that $10x = 4y$, find the ratio $x: y$.

　　A. $5:4$

　　B. $10:40$

　　C. $5:2$

　　D. $2:5$

50) What is the number of sides of a regular polygon whose interior angles are $162°$ each? (Remember, the sum of exterior angles of any polygon is $360°$).

　　A. 4

　　B. 20

　　C. 16

　　D. 18

"End of THEA Practice Test 2."

THEA Practice Test 3

Mathematics

Total Number of Questions: 50 Questions

Total time: 240 Minutes (All three sections)

You may use a non-programmable calculator for this test.

Administered *Month Year*

THEA Practice Tests

1) How many odd integers are between $\frac{-69}{6}$ and $\frac{19}{5}$?

   A. 5

   B. 11

   C. 8

   D. 9

2) Which expression correctly represents the distance between the two points shown on the number line?

   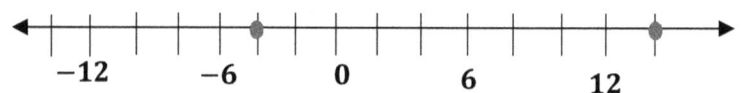

   A. $-4 - 14$

   B. $|-4 + 14|$

   C. $-4 + 14$

   D. $|4 + 14|$

3) In the triangle ABC, if angle B and angle C both equal 60°, then what is the length of side BC?

   A. 3 cm

   B. 6 cm

   C. 12 cm

   D. 4 cm

   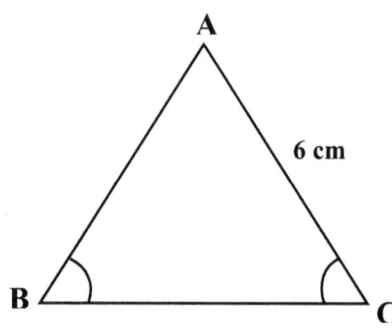

WWW.MathNotion.com

THEA Practice Tests

4) Which of the following represents the sum of the factors of 21?

   A. 32

   B. 29

   C. 11

   D. 23

5) Arrange the following fractions in order from least to greatest.

$$\frac{3}{8}, \frac{4}{7}, \frac{1}{5}, \frac{23}{25}, \frac{14}{19}$$

   A. $\frac{1}{5}, \frac{3}{8}, \frac{4}{7}, \frac{14}{19}, \frac{23}{25}$

   B. $\frac{3}{8}, \frac{4}{7}, \frac{1}{5}, \frac{14}{19}, \frac{23}{25}$

   C. $\frac{23}{25}, \frac{14}{19}, \frac{3}{8}, \frac{4}{7}, \frac{1}{5}$

   D. $\frac{14}{19}, \frac{23}{25}, \frac{1}{5}, \frac{4}{7}, \frac{3}{8}$

6) Elena earns $7.00 an hour and worked 33 hours. Her brother earns $8.25 an hour. How many hours would her brother need to work to equal Elena's earnings over 36 hours?

   A. 22.25

   B. 28

   C. 40

   D. 30.5

7) In a library, 30% of the books are fiction and the rest are non-fiction. Given that there are 1,200 more non-fiction books than fiction books, what is the total number of books in the library?

  A. 2,000

  B. 2,800

  C. 3,000

  D. 3,500

8) A map has the scale of 5 cm to 1 km. What is the actual area of a lake on ground which is represented as an area of 85 cm^2 on the map?

  A. 4.2 km^2

  B. 3.4 km^2

  C. 15 cm^2

  D. 1.5 cm^2

9) $45.45 \div 0.9 =$

  A. 50.05

  B. 50.50

  C. 5.05

  D. 5.50

THEA Practice Tests

10) What is the circumference of a circle with a radius of 11 inches?

   A. 11π

   B. 22π

   C. 121π

   D. 30.25π

11) Alfred needs to calculate his monthly water bill. His family used 28,500 gallons at a rate of $0.72 per hundred gallons. Also, there is a monthly fee of $4.40 on each period. What is his total bill?

   A. $2,310.3

   B. $205.6

   C. $209.6

   D. $29.6

12) Simplify $\dfrac{(2x^4+6x^3)}{(x^3+3x^2)} = ?$

   A. $2x$

   B. $x - 3$

   C. $3x$

   D. $3x(x+2)$

THEA Practice Tests

13) Which of the following expressions is undefined in the set of real numbers?

   A. $\sqrt[2]{144}$

   B. $\sqrt[3]{-27}$

   C. $\sqrt{-49}$

   D. $\sqrt[4]{254}$

14) If $f(x) = 3x^2$, and 5f(2a) = 180 then what could be the value of a?

   A. -3

   B. -6

   C. 6

   D. 3

15) Speed of a train is 160 miles per hour for 4 hours and 45 minutes. How many miles did the train travel?

   A. 712

   B. 625

   C. 725

   D. 655

16) Solve $9x - 7 \leq 15x + 11$

   A. $x \leq -3$

   B. $x \geq -3$

   C. $x \leq 3$

   D. $x \geq 3$

17) What is the value of 3^6?

   A. $(2 + 2)^8$

   B. 9^3

   C. $3(3^2)$

   D. $3^3 + 3^3$

18) Shane Williams puts $3,600 into a saving bank account that pays simple interest of 4.5%. How much interest will she earn after 2 years?

   A. $2,845

   B. $ 2,324

   C. $324

   D. $146

19) Three angles join to form a straight angle. One angle measure 72°. Other angle measures 34°. What is the measure of third angle?

   A. 18°

   B. 68°

   C. 106°

   D. 74°

THEA Practice Tests

20) Evaluate $\frac{21x^9 y^2 z^{-3}}{14 x^5 y^6 z^3}$.

A. $\frac{3x^4 y^3}{2 z^2}$

B. $\frac{3x^3 z^2}{2 y^2}$

C. $\frac{3x^4}{2 y^4 z^6}$

D. $\frac{7y^3}{2 x^2 z^3}$

21) Find the equation for line passing through $(2, -4)$ and $(5, 1)$.

A. $-2x - 5y = 22$

B. $3y - 5x = -22$

C. $-3x + 10y = 12$

D. $5y + 3x = -22$

22) If $x = -5$ and $y = 7$, calculate the value of $\frac{x^2+7}{y-3}$.

A. $\frac{1}{8}$

B. -8

C. 8

D. 4

THEA Practice Tests

23) The table shows the parking rates for the outside terminal area at an airport.

Ethan parked at the lot for $3\frac{1}{2}$ hours. How much did he owe?

A. $8.65

B. $7.65

C. $7.7

D. $7.85

2 hours	$3.20
Each 30 minutes after 2 hours	$1.5
24-hours Discount rate	$40

24) What is the value of x in term of c and d $\frac{c-d}{dx} = \frac{2}{7}$? (c and d>0)

A. $\frac{7}{2}(\frac{c}{d} - 1)$

B. $\frac{7}{2}(1 - \frac{c}{d})$

C. $\frac{2}{7}(\frac{c}{d} - 1)$

D. $\frac{2}{7}(1 - \frac{c}{d})$

25) Factor the equation $9x^4 - 4x^2$.

A. $2x^2(3x + 2)$

B. $x^2(2x + 3)(x - 3)$

C. $x^2(3x - 1)(x + 3)$

D. $x^2(3x - 2)(3x + 2)$

26) Which of the following are the solutions to the equation $x^2 - 14x + 48 = 0$?

   A. $-6, 8$

   B. $8, 6$

   C. $-8, -6$

   D. $6, -8$

27) Which of the following equations best represents the line in the graph below?

   A. $y = \frac{1}{5} x + 4$

   B. $y = x + 5$

   C. $y = \frac{1}{5} x - 5$

   D. $y = 5x + 1$

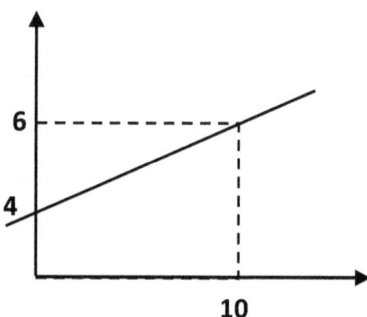

28) If $-3x + 2y = -2$ and $4x - 3y = 5$, what is the value of x?

   A. 7

   B. 1

   C. -2

   D. -4

29) lengths of two sides of a triangle are 8 and 5. Which of the following could Not be the measure of third side?

   A. 2

   B. 4

   C. 10

   D. 12

30) The following data set is given: 111, 140, 127, 165, 159, 120.

Adding which number to the set will increase its mean?

A. 134

B. 129

C. 132

D. 148

31) An award for best education improvement is awarded annually to a winning US state, and the winners from 2002 to 2011 are given in the table below. Find the mode of this set of states.

Year	2002	2003	2004	2005	2006	2007	2008	2009	2010	2011
State	New Jersey	New York	Oregon	New York	Oregon	California	Oregon	Ohio	Oregon	New York

A. Ohio and Oregon

B. California and Ohio

C. New York and Oregon

D. New Jersey and California

32) How long is a distance of 8 km if measured on a map with a scale of 1:50,000?

A. 14

B. 16

C. 8

D. 3

33) What is the area of a square, if its side measures $\sqrt{13}$ m?

A. $\sqrt{13}$

B. $3\sqrt{13}$

C. $4\sqrt{13}$

D. 13

34) What is the answer of the following equation $2x^4y^4 (3x^3y^3)^4 =?$

A. $81x^{11}y^{11}$

B. $162x^{16}y^{16}$

C. $46x^{10}y^8$

D. $112x^8y^8$

35) The circle graph below shows the type of pizza that people prefer for lunch. If 220 people were surveyed, how many people preferred Margherita?

A. 65

B. 50

C. 55

D. 20

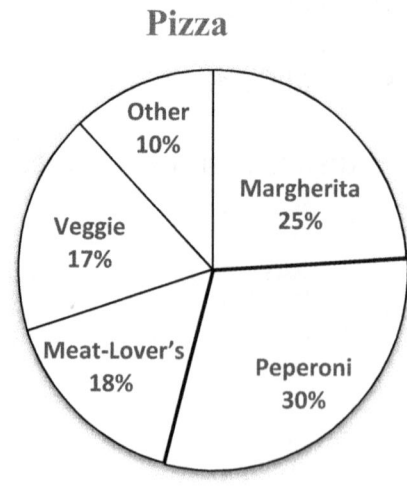

THEA Practice Tests

36) Rosie is x years old. She is 3 years older than her twin brothers Milan and Marcel. What is the mean age of the three children?

   A. $x + 2$

   B. $x - 3$

   C. $x + 3$

   D. $x - 2$

37) a is inversely proportional to $(2b - 5)$. If a = 8 and b = 9, express a in terms of b.

   A. $a = 2b - 18$

   B. $a = 108(2b - 5)$

   C. $a = \frac{104}{2b-5}$

   D. $a = \frac{2b-5}{104}$

38) A baseball has a volume of 36π. What is the length of the diameter?

   A. 15

   B. 6

   C. 3

   D. 12

39) 154 is What percent of 140?

   A. 110 %

   B. 60 %

   C. 120 %

   D. 220 %

40) A phone manufacturer makes 18,000 phone a year. The company randomly selects 600 of the phones to sample for inspection. The company discovers that there are 5 faulty phones in the sample. Based on the sample, how many of the 18,000 total phones are likely to be faulty?

   A. 300

   B. 50

   C. 250

   D. 150

41) Emma and Mia buy a total of 28 books. Emma bought 4 more books than Mia did. How many books did Emma buy?

   A. 24

   B. 16

   C. 12

   D. 8

42) A bag contains 11 white balls, 6 red balls and 7 black balls. A ball is picked from the bag at random. Find the probability of picking a red ball?

   A. 1.25

   B. 2.25

   C. 0.45

   D. 0.25

43) The radius of the following cylinder is 4 inches, and its height are 9 inches. What is the surface area of the cylinder in square inches? (π=3.14)

A. 326.56

B. 366.5

C. 5.56

D. 36.65

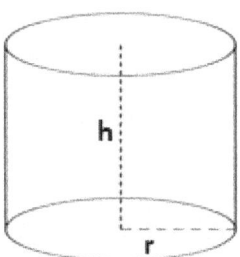

44) The line n has a slope of $\frac{a}{b}$, where a and b are integers. What is the slope of a line that is perpendicular to line n?

A. $-\frac{a}{b}$

B. $\frac{a}{b}$

C. $\frac{b}{a}$

D. $-\frac{b}{a}$

45) In a store, 38% of customers are female. If the total number of customers is 750, then how many male customers are dealing with the store?

A. 421

B. 726

C. 465

D. 356

46) If 66.33 kg is divided into two parts, in a ratio of 8:3, how many kg is the smaller share?

   A. 48.24 kg

   B. 18.09 kg

   C. 1.89 kg

   D. 19.09 kg

47) Express as a single fraction in its simplest form: $\frac{5}{(x+2)} - \frac{2}{(4x-1)} = ?$

   A. $\frac{18x-9}{(x+2)(4x-1)}$

   B. $\frac{9x-18}{(x+2)(4x-1)}$

   C. $\frac{15x-17}{4x+2}$

   D. $\frac{-17}{4x-1}$

48) Ryan is x years old, and her sister Mitzi is $(6x - 20)$ years old. Given that Mitzi is twice as old as Ryan, what is Mitzi's age?

   A. 7

   B. 11

   C. 10

   D. 14

THEA Practice Tests

49) The set of possible values of p is {3, 5, 11}. What is the set of possible values of h if $2h = 5p + 1$?

   A. {6,17,25}

   B. {8,14,26}

   C. {8, 13, 28}

   D. {6,11,28}

50) In the infinitely repeating decimal below, 4 is the second digit in the repeating pattern. What is the 764th digit? $\frac{1}{21} = 0.\overline{047619}$

   A. 4

   B. 7

   C. 9

   D. 0

"End of THEA Practice Test 3."

THEA Practice Test 4

Mathematics

Total Number of Questions: 50 Questions

Total time: 240 Minutes (All three sections)

You may use a non-programmable calculator for this test.

Administered *Month Year*

THEA Practice Tests

1) What is the sum of the smallest prime number and five times the largest negative even integer?

   A. −10

   B. 14

   C. −8

   D. −12

2) Sam's incomes and expenditures for the first season of the last year are given in the table below. In which month were her savings the highest?

   A. January

   B. March

   C. February

   D. All months were the same.

Month	Income	Cost
January	$3,740	$2,202
February	$3,855	$1,985
March	$3,970	$2,005

3) What is the number a, if the result of adding a to 29 is the same as subtracting $4a$ from 244?

   A. −59

   B. 34

   C. 122

   D. 43

THEA Practice Tests

4) Solve these fractions and reduce to its simplest terms: $8\frac{7}{24} - 6\frac{2}{3} + 2\frac{5}{6} =$

   A. $4\frac{11}{24}$

   B. $-4\frac{7}{24}$

   C. $\frac{7}{8}$

   D. $5\frac{3}{8}$

5) Calculate, $3^3 + 3 + 3^0 = ?$

   A. 3^4

   B. 30

   C. 31

   D. 35

6) Which decimal is equivalent to $\frac{153}{170}$?

   A. 0.109

   B. 1.09

   C. 0.09

   D. 0.9

7) The price of a shirt increased from $30 to $32.10. What is the percentage increase in the price?

   A. 7%

   B. 0.7%

   C. 0.93%

   D. 1.7%

8) Find the solution set of the following equation: $|2x - 3| = 7$

   A. $\{5, -2\}$

   B. $\{2, -5\}$

   C. $\{2\}$

   D. $\{-5, -2\}$

9) What is the solution to the pair of equations below? $\begin{cases} 2x - 5y = 16 \\ 2x + y = 4 \end{cases}$

   A. $x = 0$ and $y = -3$

   B. $x = -3$ and $y = 0$

   C. $x = 3$ and $y = -2$

   D. $x = 3$ and $y = -3$

10) The train each 30 minutes passes an average 5 stations. At this rate, how many stations will it pass in three hours.

   A. 20

   B. 30

   C. 45

   D. 60

THEA Practice Tests

11) If Sofia buys a shirt marked down 16 percent from its $160 while Natalia buys the same shirt mark down only 11 percent, how much more does Natalia pay for the shirt?

  A. $80

  B. $4.8

  C. $16

  D. $8

12) Find the circumference of the circle in terms of π.

  A. 9π in.

  B. 72π in.

  C. 36 π in.

  D. 112π in.

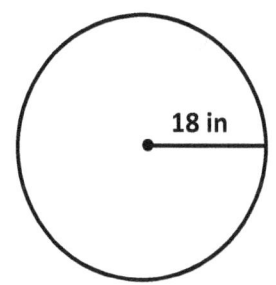

13) Find the volume of rectangular prism below?

  A. 1,880 in^3

  B. 1,080 in^3

  C. 2,280 in^3

  D. 285 in^3

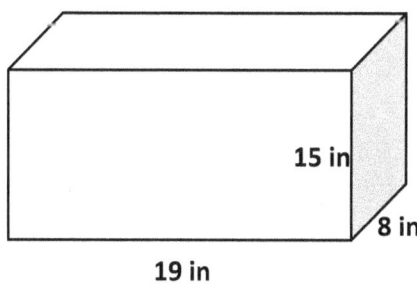

14) What is $\sqrt[4]{5^{-8}}$ in simplest form?

A. $\dfrac{1}{2,250}$

B. $\dfrac{1}{125}$

C. $\dfrac{1}{5}$

D. $\dfrac{1}{25}$

15) Which statement correctly describes the value of N in the equation below?

$7(4N - 10) = 4(7N - 15)$?

A. N has no correct solutions.

B. N=1 is one solution.

C. N has infinitely many correct solutions.

D. N=0 is one solution.

16) What is the maximum amount of grain, the silo can hold, in cubic feet?

A. $1,620\pi\ m^3$

B. $405\pi\ m^3$

C. $810\pi\ m^3$

D. $1,200\pi\ m^3$

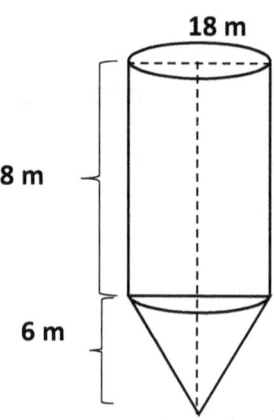

THEA Practice Tests

17) What is 7.25×10^{-5} in standard form?

   A. $-72,500$

   B. -0.0000725

   C. $\dfrac{1}{725,000}$

   D. 0.0000725

18) What is the value of x in the triangle?

   A. $64°$

   B. $130°$

   C. $50°$

   D. $60°$

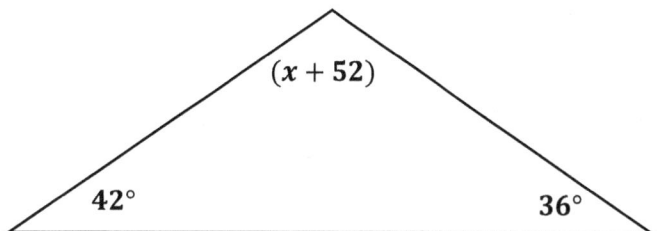

19) Find the length of the unknown side.

   A. 22 ft

   B. 20 ft

   C. 18 ft

   D. 12 ft

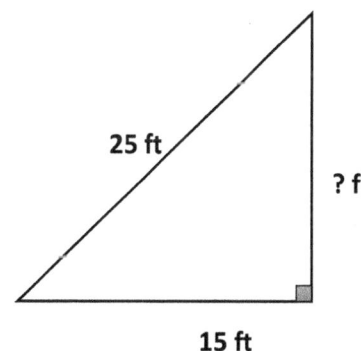

20) A store sells all of its products at a price 12% greater than the price the store paid for the product. How much does the store sell a product if the store paid $150 for it?

A. $180

B. $160

C. $168

D. $38

21) What is the area of shaded region?

A. 95

B. 35

C. 130

D. 165

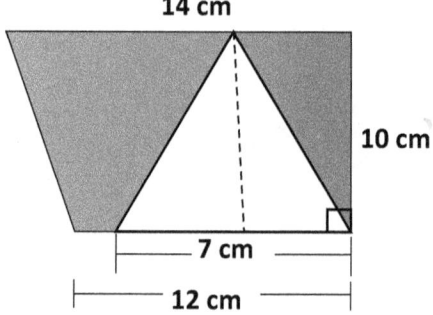

22) if $xy - 8x = 54$ and $y - 8 = 6$, then $x =$?

A. 8

B. 14

C. 6

D. 9

23) Which is the value of x in the equation $\frac{x}{4} = x - 6$?

   A. 4

   B. 6

   C. 8

   D. 12

24) What is the value of x, If $-5x + 4y = 7$ and $-4x + 3y = 5$?

   A. -2

   B. 0

   C. 3

   D. 1

25) What is the probability of Not spinning at F?

   A. $\frac{5}{8}$

   B. $\frac{3}{8}$

   C. $\frac{1}{8}$

   D. $\frac{7}{8}$

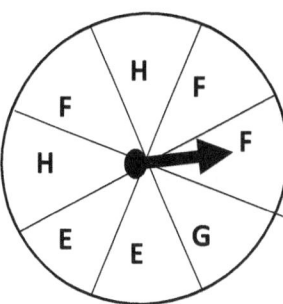

26) Ella bought 32 movies for $5.04 per movie. Which equation shows the BEST estimate of the total cost?

   A. 30 × $6 = $180

   B. 30 × $5 = $150

   C. 32 × $5 = $160

   D. 32 × $6 = $192

THEA Practice Tests

27) A pick-up truck travels 50 mile on 7 L of gasoline when driven on a smooth road. If the cost of gasoline is $1.40/L, which is the cost of 1,500 mile of highway (smooth)?

   A. $141

   B. $1,249

   C. $129

   D. $294

28) A position of subway station and Grace's house shown by a grid. The station is located at $(-6, -9)$, and her house is located at $(2, -3)$. What is the distance between her house and the subway stop?

   A. 6

   B. 10

   C. $10\sqrt{2}$

   D. 15

29) For the following set of numbers find the median.

   225, 75, 280, 89, 198, 124, 512.

   A. 198

   B. 124

   C. 225

   D. 89

THEA Practice Tests

30) What is solution to the equation $\sqrt{4x-7} = 9$?

   A. -18

   B. -14

   C. 10

   D. 22

31) The equation $x = 4y - 8$ has a y-intercept of?

   A. 2

   B. -2

   C. $\frac{1}{2}$

   D. $-\frac{1}{2}$

32) If $3^{2x} = 729$, then $x = ?$

   A. 4

   B. 6

   C. 3

   D. 2

33) Which is the smallest positive integer which is divisible by both 18 and 64.

   A. 108

   B. 54

   C. 576

   D. 124

THEA Practice Tests

34) A cube has total surface area of 96 cm². what is the volume of the cube in cm³?

   A. 8

   B. 64

   C. 36

   D. 108

35) Which is the value of x^2, if $x^2 + 2x = 35$?

   A. 12

   B. -7

   C. 36

   D. 25

36) Each of 5 pitchers can contain up to $\frac{3}{5}$ L of water. If each of the pitcher is at least the half full, which of the following expressions represents the total amount of water, W, contained on all 5 pitchers?

   A. $2.5 < w < 5$

   B. $1.5 < w < 5$

   C. $0.5 < w < 3$

   D. $1.5 < w < 3$

37) Ages of the players on a volleyball team is given. Which is the range of their ages? 47, 32, 17, 65, 22, 40, 28, 57, 20, 18, 56, 39

   A. 38

   B. 19

   C. 48

   D. 52

38) Find the area of the circle to the nearest tenth. Use 3.14 for π.

   A. 201

   B. 202

   C. 200.9

   D. 109

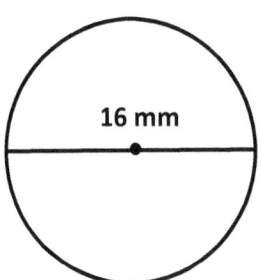

39) What is the simplest form of the expression $\dfrac{2x^2-15x+7}{4(x^2-\tfrac{1}{4})}$?

   A. $\dfrac{x+7}{2(x-\tfrac{1}{2})}$

   B. $\dfrac{x-7}{2x+1}$

   C. $\dfrac{x+7}{2x+1}$

   D. $\dfrac{x-7}{x-2}$

THEA Practice Tests

40) What is the value of $\left(\frac{1}{3}\right)^{-4}$?

A. $\frac{1}{81}$

B. -81

C. $-\frac{1}{81}$

D. 81

41) What is the simplest form of the expression $\frac{(5x^{-3}y^4)^3}{100y^{-2}z^2}$, (using positive exponent)?

A. $\frac{4y^{14}z}{59}$

B. $\frac{5y^{14}}{4x^9z^2}$

C. $\frac{5x^8z^2}{4zy^8}$

D. $\frac{x^{14}}{y^9}$

42) Some fruit sells for $30 per kilograms. What is the price in cent per gram?

A. 0.003

B. 0.3

C. 0.03

D. 3

THEA Practice Tests

43) The price of water triples every 5 years. If the price of water on January 1st, 2012, is $6 per gallon, what is the equation that would be used to calculate the price(P) of water on January 1st, 2007?

   A. $6P = 6$

   B. $\frac{P}{6} = 3$

   C. $3p = 6$

   D. $6P = 3$

44) The line $4y + 5 = 20x + 9$ and $5y + 4 = x + 7$ are?

   A. Perpendicular

   B. Parallel

   C. The same line

   D. Neither parallel nor perpendicular

45) A rectangular box measures $5\frac{1}{3}$ feet by $8\frac{2}{5}$ feet. It is divided into four equal parts. What is the area of one of those parts?

   A. 11

   B. 9.2

   C. 11.2

   D. 22.4

46) Find the slope of the line.

A. $\dfrac{1}{2}$

B. $\dfrac{1}{3}$

C. 2

D. $-\dfrac{1}{4}$

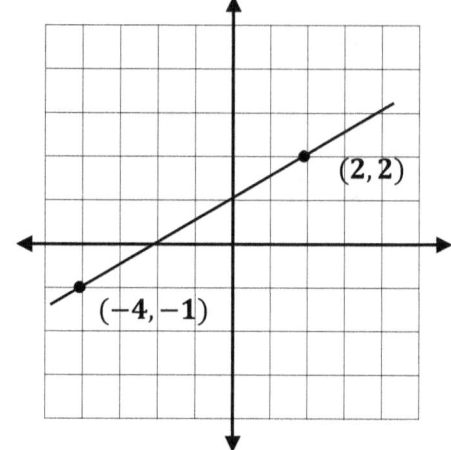

47) Amelia cuts a piece of birthday cake as shown below. What is the volume of the piece of cake?

A. 425 cm^3

B. 375 cm^3

C. 750 cm^3

D. 250 cm^3

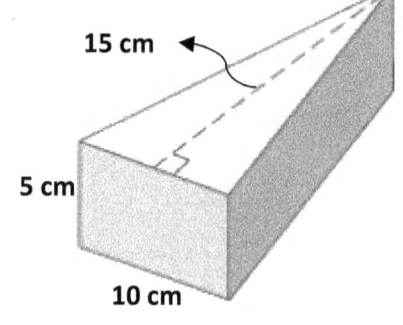

48) Let $f(x) = 4x - 7$. If $f(a) = -19$ and $f(b) = 9$, then what is $f(a + b)$?

A. 3

B. 5

C. -3

D. -5

49) Given that $28x = 12y$, find the ratio $x:y$.

   A. $5:9$

   B. $4:11$

   C. $6:5$

   D. $3:7$

50) What is the number of sides of a regular polygon whose interior angles are $168°$ each? (Remember, the sum of exterior angles of any polygon is $360°$).

   A. 6

   B. 30

   C. 22

   D. 12

"End of THEA Practice Test 4."

THEA Practice Test 5

Mathematics

Total Number of Questions: 50 Questions

Total time: 240 Minutes (All three sections)

You may use a non-programmable calculator for this test.

Administered Month Year

THEA Practice Tests

1) How many even integers are between $\frac{-26}{3}$ and $\frac{75}{8}$?

   A. 8

   B. 5

   C. 9

   D. 11

2) Which expression correctly represents the distance between the two points shown on the number line?

   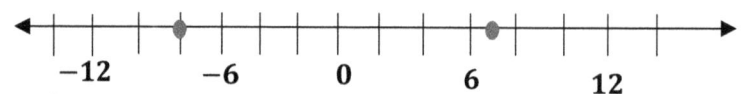

   A. $-8 - 7$

   B. $|-8 + 7|$

   C. $-8 + 7$

   D. $|8 + 7|$

3) In the triangle ABC, if angle B and angle C both equal 60°, then what is the length of side BC?

   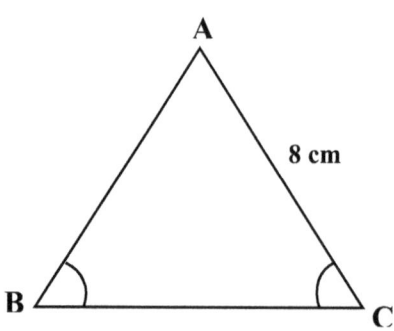

   A. 4 cm

   B. 8 cm

   C. 12 cm

   D. 16 cm

WWW.MathNotion.com

THEA Practice Tests

4) Which equation can be equal "4 more than the ratio of a number to 5 is equal to 7 less than the number"?

   A. $4x - 5 = 7 - x$

   B. $4 + \frac{x}{5} = x - 7$

   C. $\frac{4}{5}x - 7 = 5x$

   D. $4 + 5x = 7 - x$

5) Arrange the following fractions in order from least to greatest.

$$\frac{3}{7}, \frac{5}{9}, \frac{1}{3}, \frac{19}{21}, \frac{11}{18}$$

   A. $\frac{1}{3}, \frac{3}{7}, \frac{5}{9}, \frac{11}{18}, \frac{19}{21}$

   B. $\frac{5}{9}, \frac{3}{7}, \frac{1}{3}, \frac{11}{18}, \frac{19}{21}$

   C. $\frac{19}{21}, \frac{11}{18}, \frac{5}{9}, \frac{3}{7}, \frac{1}{3}$

   D. $\frac{11}{18}, \frac{19}{21}, \frac{1}{3}, \frac{3}{7}, \frac{5}{9}$

6) Elena earns $9.20 an hour and worked 35 hours. Her brother earns $11.50 an hour. How many hours would her brother need to work to equal Elena's earnings over 40 hours?

   A. 15.22

   B. 28

   C. 35

   D. 80.50

7) In a library, 40% of the books are fiction and the rest are non-fiction. Given that there are 1,200 more non-fiction books than fiction books, what is the total number of books in the library?

A. 2,000

B. 5,000

C. 6,000

D. 4,000

8) A map has the scale of 5 cm to 1 km. What is the actual area of a lake on ground which is represented as an area of 80 cm^2 on the map?

A. 16 km^2

B. 3.2 km^2

C. 16 cm^2

D. 3.2 cm^2

9) $12.24 \div 0.4 =$

A. 3.06

B. 30.60

C. 30.06

D. 3.006

THEA Practice Tests

10) What is the circumference of a circle with a radius of 8 inches?

   A. 64π

   B. 16π

   C. 128π

   D. 48π

11) Alfred needs to calculate his monthly water bill. His family used 23,700 gallons at a rate of $0.95 per hundred gallons. Also, there is a monthly fee of $4.20 on each period. What is his total bill?

   A. $22,519.20

   B. $225.15

   C. $229.35

   D. $6.45

12) Simplify $\frac{(x^3-x^2)}{(x^2-x)} = ?$

   A. x

   B. $x - 1$

   C. $2x$

   D. $x(x-1)$

13) Which of the following expressions is undefined in the set of real numbers?

   A. $\sqrt[2]{148}$

   B. $\sqrt[3]{-27}$

   C. $\sqrt{-81}$

   D. $\sqrt[4]{16}$

14) If $f(x) = 5x^2$, and 3f(2a) = 540 then what could be the value of a?

   A. -3

   B. -1

   C. 1

   D. 3

15) Speed of a train is 110 mile per hour for 2 hours and 30 minutes. How many miles did the train travel?

   A. 275

   B. 256

   C. 253

   D. 270

16) Solve $10x - 7 \leq 6x + 5$

   A. $x \leq -3$

   B. $x \geq -3$

   C. $x \leq 3$

   D. $x \geq 3$

THEA Practice Tests

17) What is the value of 3^4?

   A. $(3+3)^2$

   B. 9^2

   C. $3(3^2)$

   D. $3^2 + 3^2$

18) Shane Williams puts $3,800 into a saving bank account that pays simple interest of 4.5%. How much interest will she earn after 3 years?

   A. $5,130

   B. $ 1,710

   C. $513

   D. $171

19) Three angles join to form a straight angle. One angle measure 55°. Other angle measures 30°. What is the measure of third angle?

   A. 5°

   B. 15°

   C. 35°

   D. 95°

THEA Practice Tests

20) Evaluate $\frac{32x^5y^7z^{-2}}{12x^2y^9z^0}$.

A. $\frac{4x^3y^2}{3z^2}$

B. $\frac{4x^3z^2}{3y^2}$

C. $\frac{8x^3}{3y^2z^2}$

D. $\frac{8y^2}{3x^3z^2}$

21) Find the equation for line passing through $(2, -1)$ and $(4, 2)$.

A. $-3x - 2y = 10$

B. $2y - 3x = -8$

C. $-3x + 2y = 10$

D. $2y + 3x = -8$

22) If $x = -2$ and $y = 2$, calculate the value of $\frac{x^2+5}{y+1}$.

A. $\frac{1}{3}$

B. -3

C. 3

D. 2

THEA Practice Tests

23) The table shows the parking rates for the outside terminal area at an airport. Ethan parked at the lot for $3\frac{1}{2}$ hours. How much did he owe?

A. $4:25

B. $3:50

C. $5:75

D. $6:50

2 hours	$3.5
Each 30 minutes after 2 hours	$0.75
24-hours Discount rate	$40

24) What is the value of x in term of c and d $\frac{c-d}{dx} = \frac{2}{3}$? (c and d>0)

A. $\frac{3}{2}(\frac{c}{d} - 1)$

B. $\frac{3}{2}(1 - \frac{c}{d})$

C. $\frac{2}{3}(\frac{c}{d} - 1)$

D. $\frac{2}{3}(1 - \frac{c}{d})$

25) Factor the equation $x^3 - x$.

A. $3x(x - 1)$

B. $x(x + 2)(x - 1)$

C. $x(x - 2)(x + 1)$

D. $x(x - 1)(x + 1)$

WWW.MathNotion.com

26) Which of the following are the solutions to the equation $x^2 - 10x + 24 = 0$?

   A. $-12, 2$

   B. $6, 4$

   C. $-4, -6$

   D. $12, -2$

27) Which of the following equations best represents the line in the graph below?

   A. $y = \frac{1}{2}x + 2$

   B. $y = x + 5$

   C. $y = \frac{1}{2}x - 2$

   D. $y = x + 2$

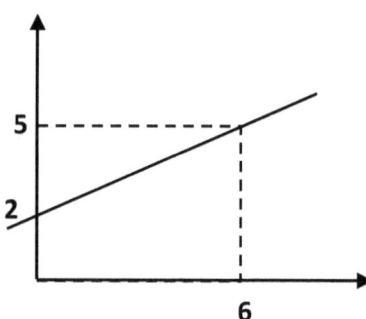

28) If $-3x + 4y = -5$ and $2x - 5y = 8$, what is the value of x?

   A. 2

   B. 2.60

   C. -8

   D. -1

29) lengths of Two sides of a triangle are 5 and 8. Which of the following could Not be the measure of third side?

   A. 2

   B. 4

   C. 9

   D. 11

THEA Practice Tests

30) The following data set is given: 134, 118, 148, 184, 159, 151

    Adding which number to the set will increase its mean?

    A. 129

    B. 139

    C. 149

    D. 159

31) An award for best education improvement is awarded annually to a winning US state, and the winners from 2002 to 2011 are given in the table below. Find the mode of this set of states.

Year	2002	2003	2004	2005	2006	2007	2008	2009	2010	2011
State	New York	Ohio	New Jersey	New York	Ohio	California	Ohio	Oregon	New York	New Jersey

A. New York and New Jersey

B. California and Oregon

C. Ohio and New York

D. Ohio and California

32) How long is a distance of 6 km if measured on a map with a scale of 1:50,000?

    A. 6

    B. 12

    C. 16

    D. 18

33) What is the area of a square, if its side measures $\sqrt{5}$ m?

   A. $\sqrt{5}$

   B. $5\sqrt{5}$

   C. $2\sqrt{5}$

   D. 5

34) What is the answer of the following equation $2x^4 y^3 (4x^3 y^0)^3 = ?$

   A. $24x^{10} y^6$

   B. $128x^{13} y^3$

   C. $128x^{13} y^6$

   D. $24x^{13} y^3$

35) The circle graph below shows the type of pizza that people prefer for lunch. If 250 people were surveyed, how many people preferred Meat lovers?

   A. 18

   B. 25

   C. 45

   D. 55

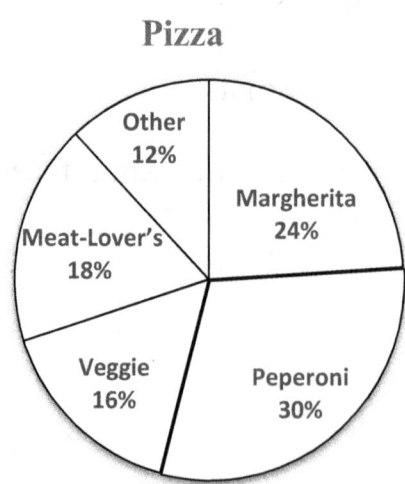

36) Rosie is *x* years old. She is 3 years older than her twin brothers Milan and Marcel. What is the mean age of the three children?

   A. $x + 2$

   B. $x - 3$

   C. $x + 3$

   D. $x - 2$

37) a is inversely proportional to $(2b - 5)$. If a = 15 and b = 7, express a in terms of b.

   A. $a = 2b - 5$

   B. $a = 105(2b - 5)$

   C. $a = \frac{135}{2b-5}$

   D. $a = \frac{2b-5}{135}$

38) A baseball has a volume of 36π. What is the length of the diameter?

   A. 3

   B. 6

   C. 9

   D. 12

39) 221 is What percent of 170?

   A. 130 %

   B. 77 %

   C. 30 %

   D. 123 %

40) A phone manufacturer makes 12,000 phone a year. The company randomly selects 200 of the phones to sample for inspection. The company discovers that there are 3 faulty phones in the sample. Based on the sample, how many of the 12,000 total phones are likely to be faulty?

A. 20

B. 60

C. 120

D. 180

41) Emma and Mia buy a total of 17 books. Emma bought 5 more books than Mia did. How many books did Emma buy?

A. 7

B. 11

C. 12

D. 15

42) A bag contains 5 white balls, 4 red balls and 7 black balls. A ball is picked from the bag at random. Find the probability of picking a red ball?

A. 1.25

B. 0.28

C. 0.33

D. 0.25

43) The radius of the following cylinder is 6 inches, and its height are 9 inches. What is the surface area of the cylinder in square inches? (π=3.14)

A. 565.2

B. 109.9

C. 1,130

D. 94.2

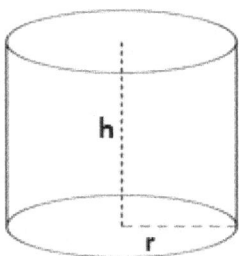

44) The line n has a slope of $\frac{c}{d}$, where c and d are integers. What is the slope of a line that is perpendicular to line n?

A. $\frac{c}{d}$

B. $-\frac{c}{d}$

C. $\frac{d}{c}$

D. $-\frac{d}{c}$

45) In a store, 38% of customers are female. If the total number of customers is 950, then how many male customers are dealing with the store?

A. 228

B. 361

C. 589

D. 675

THEA Practice Tests

46) If 54.5 kg is divided into two parts, in a ratio of 8:3, how many kg is the smaller share?

   A. 18.17 kg

   B. 14.86 kg

   C. 2.67 kg

   D. 6.81 kg

47) Express as a single fraction in its simplest form: $\dfrac{5}{(x-2)} - \dfrac{6}{(3x+1)}$

   A. $\dfrac{9x+17}{(x-2)(3x+1)}$

   B. $\dfrac{9x-17}{(x-2)(3x+1)}$

   C. $\dfrac{9x+17}{3x-2}$

   D. $\dfrac{-1}{3x-2}$

48) Ryan is x years old, and her sister Mitzi is $(5x - 18)$ years old. Given that Mitzi is twice as old as Ryan, what is Mitzi's age?

   A. 6

   B. 9

   C. 12

   D. 18

THEA Practice Tests

49) The set of possible values of p is {2,5,11}. What is the set of possible values of h if $3h = 2p + 2$?

   A. {2,5,11}

   B. {6,15,33}

   C. {2,4,8}

   D. {6,12,24}

50) In the infinitely repeating decimal below, 1 is the first digit in the repeating pattern. What is the 391st digit? $\frac{1}{7} = \overline{0.142857}$

   A. 1

   B. 4

   C. 8

   D. 5

"End of THEA Practice Test 5."

THEA Practice Test 6

Mathematics

Total Number of Questions: 50 Questions

Total time: 240 Minutes (All three sections)

You may use a non-programmable calculator for this test.

Administered *Month Year*

THEA Practice Tests

1) What is the sum of the smallest prime number and three times the largest negative even integer?

   A. −2

   B. 0

   C. −4

   D. −6

2) Sam's incomes and expenditures for the first season of the last year are given in the table below. In which month were her savings the highest?

   A. January

   B. February

   C. March

   D. All months were the same.

Month	Income	Cost
January	$3,025	$1,870
February	$3,405	$1,916
March	$3,280	$1,962

3) What is the number a, if the result of adding a to 42 is the same as subtracting $3a$ from 230?

   A. −94

   B. 68

   C. 188

   D. 47

THEA Practice Tests

4) Solve these fractions and reduce to its simplest terms: $4\frac{1}{21} - 5\frac{4}{7} + 2\frac{1}{3} =$

   A. $\frac{17}{21}$

   B. $-1\frac{4}{21}$

   C. $\frac{6}{7}$

   D. $1\frac{2}{3}$

5) Calculate, $3^3 + 3^2 + 3 = ?$

   A. 9^6

   B. 18

   C. 39

   D. 729

6) Which decimal is equivalent to $\frac{108}{270}$?

   A. 0.108

   B. 0.27

   C. 0.04

   D. 0.4

7) The price of a shirt increased from $30 to $31.50. What is the percentage increase in the price?

   A. 5%

   B. 0.5%

   C. 0.95%

   D. 1.5%

THEA Practice Tests

8) Find the solution set of the following equation: $|2x - 3| = 5$

   A. $\{4, -1\}$

   B. $\{1, -1\}$

   C. $\{5\}$

   D. $\{-4, 1\}$

9) What is the solution to the pair of equations below? $\begin{cases} x - 3y = 1 \\ 2x + y = 2 \end{cases}$

   A. $x = 4$ and $y = 1$

   B. $x = 0$ and $y = 1$

   C. $x = 1$ and $y = 0$

   D. $x = 1$ and $y = -3$

10) The train each 15 minutes passes an average 4 stations. At this rate, how many stations will it pass in two hours.

    A. 8

    B. 32

    C. 16

    D. 48

THEA Practice Tests

11) If Sofia buys a shirt marked down 8 percent from its $180 while Natalia buys the same shirt mark down only 6 percent, how much more does Natalia pay for the shirt?

   A. $0.36

   B. $1.80

   C. $2

   D. $3.60

12) Find the circumference of the circle in terms of π.

   A. 21.5π in.

   B. 43π in.

   C. 86π in.

   D. $1,849\pi$ in.

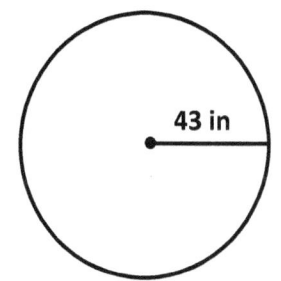

13) Find the volume of rectangular prism below?

   A. $513\ in^3$

   B. $1026\ in^3$

   C. $2,052\ in^3$

   D. $4,104\ in^3$

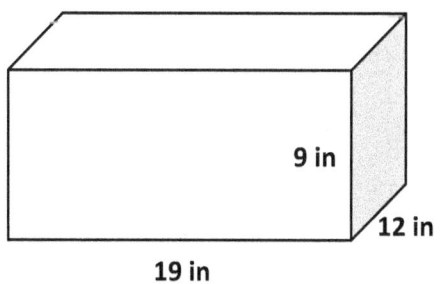

THEA Practice Tests

14) What is $\sqrt[4]{4^{-8}}$ in simplest form?

A. $\dfrac{1}{65,536}$

B. $\dfrac{1}{256}$

C. $\dfrac{1}{64}$

D. $\dfrac{1}{16}$

15) Which statement correctly describes the value of N in the equation below?

$4(7N - 12) = 7(4N - 12)$?

A. N has no correct solutions.

B. N=0 is one solution.

C. N has infinitely many correct solutions.

D. N=1 is one solution.

16) What is the maximum amount of grain, the silo can hold, in cubic feet?

A. $144\pi \; m^3$

B. $288\pi \; m^3$

C. $896\pi \; m^3$

D. $1,152\pi \; m^3$

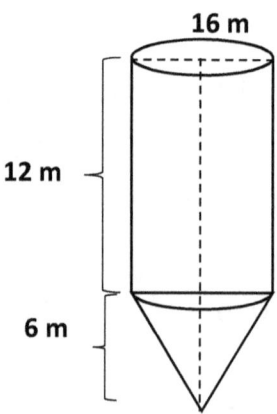

THEA Practice Tests

17) What is 3.21×10^{-4} in standard form?

   A. −32,120

   B. −0.000321

   C. $\frac{1}{32,100}$

   D. 0.000321

18) What is the value of x in the triangle?

   A. 77°

   B. 130°

   C. 50°

   D. 103°

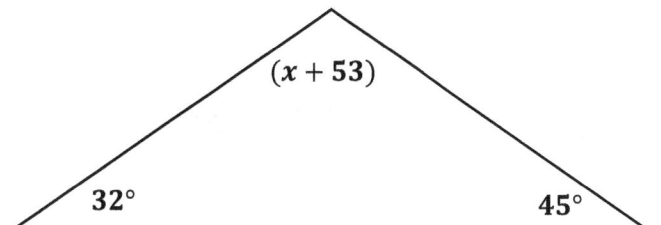

19) Find the length of the unknown side.

   A. 23.3 ft

   B. 16 ft

   C. 256 ft

   D. 8 ft

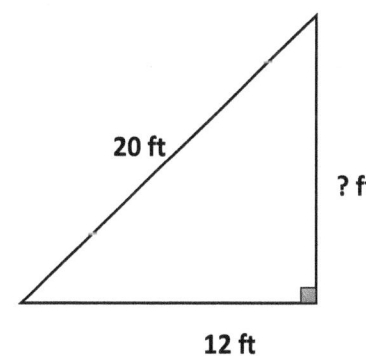

20) A store sells all of its products at a price 18% greater than the price the store paid for the product. How much does the store sell a product if the store paid $250 for it?

A. $268

B. $205

C. $295

D. $45

21) What is the area of shaded region?

A. 81

B. 54

C. 27

D. 18

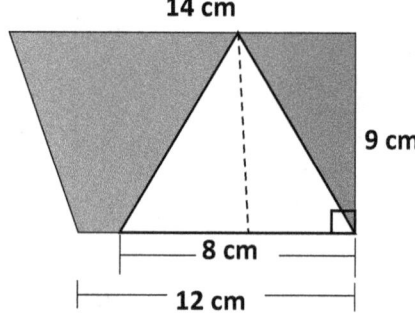

22) if $xy - 5x = 32$ and $y - 5 = 8$, then $x =$?

A. 32

B. 16

C. 8

D. 4

23) Which is the value of x in the equation $\frac{x}{3} = x - 4$?

   A. 1

   B. 2

   C. 3

   D. 6

24) What is the value of x, If $-5x + 2y = 20$ and $-4x + 3y = 9$?

   A. -3

   B. -4

   C. -5

   D. -6

25) What is the probability of Not spinning at F?

   A. $\frac{5}{8}$

   B. $\frac{3}{8}$

   C. $\frac{1}{3}$

   D. $\frac{1}{5}$

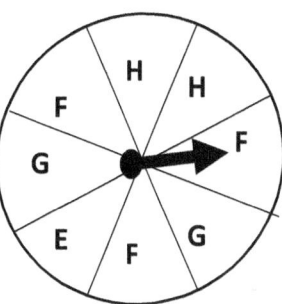

26) Ella bought 23 movies for $2.08 per movie. Which equation shows the BEST estimate of the total cost?

   A. 20 × $2 = $40

   B. 20 × $3 = $60

   C. 23 × $2 = $46

   D. 23 × $3 = $69

27) A pick-up truck travels 75 mile on 12 L of gasoline when driven on a smooth road. If the cost of gasoline is $1.05/L, which is the cost of 1,700 mile of highway (smooth)?

A. $11,156.25

B. $1,785

C. $148.75

D. $285.60

28) A position of subway station and Grace's house shown by a grid. The station is located at $(-2, -7)$, and her house is located at $(6, -1)$. What is the distance between her house and the subway stop?

A. 8

B. 10

C. $8\sqrt{2}$

D. 6

29) For the following set of numbers find the median.

212, 122, 318, 122, 186, 165, 334.

A. 186

B. 122

C. 154

D. 342

THEA Practice Tests

30) What is solution to the equation $\sqrt{3x-1} = 4$?

   A. -1

   B. -5

   C. 1

   D. 5

31) The equation $x = 2y - 4$ has a y-intercept of?

   A. 2

   B. -4

   C. $\frac{1}{2}$

   D. $-\frac{1}{4}$

32) If $3^{2x} = 81$, then $x = ?$

   A. 4

   B. 3

   C. 2

   D. 1

33) Which is the smallest positive integer which is divisible by both 24 and 30.

   A. 6

   B. 60

   C. 120

   D. 240

THEA Practice Tests

34) A cube has total surface area of 54 cm². what is the volume of the cube in cm³?

   A. 81

   B. 27

   C. 9

   D. 3

35) Which is the value of x^2, if $x^2 + x = 30$?

   A. 16

   B. -25

   C. 30

   D. 36

36) Each of 5 pitchers can contain up to $\frac{3}{5}$ L of water. If each of the pitcher is at least the half full, which of the following expressions represents the total amount of water, W, contained on all 5 pitchers?

   A. $0.6 < w < 6$

   B. $0 < w < 1.5$

   C. $0 < w < 3$

   D. $1.5 < w < 3$

37) Ages of the players on a volleyball team is given. Which is the range of their ages? 34, 41, 35, 45, 32, 34, 31, 29, 40, 28, 37, 39

A. 5

B. 11

C. 17

D. 45

38) Find the area of the circle to the nearest tenth. Use 3.14 for π.

A. 153.9

B. 615.4

C. 44.0

D. 22

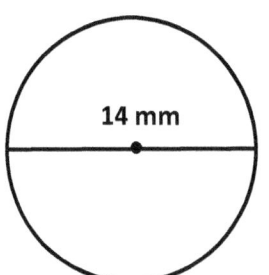

39) What is the simplest form of the expression $\frac{2x^2-7x-4}{4(x^2-\frac{1}{4})}$?

A. $\frac{x+4}{4(x-\frac{1}{2})}$

B. $\frac{x-4}{2x-1}$

C. $\frac{x+4}{2x+1}$

D. $\frac{2x-1}{x-4}$

THEA Practice Tests

40) What is the value of $\left(\frac{1}{3}\right)^{-3}$?

   A. $\frac{1}{27}$

   B. -27

   C. $-\frac{1}{27}$

   D. 27

41) What is the simplest form of the expression $\frac{(5x^{-2}y^3)^2}{125y^{-2}z^{-1}}$, (using positive exponent)?

   A. $\frac{2y^8z^2}{25x^4}$

   B. $\frac{y^8z}{5x^4}$

   C. $\frac{2x^4}{25zy^8}$

   D. $\frac{x^4}{5y^8z}$

42) Some fruit sells for $10 per kilograms. What is the price in cent per gram?

   A. 0.001

   B. 0.1

   C. 0.01

   D. 1

43) The price of water doubles every 5 years. If the price of water on January 1st, 2012, is $2 per gallon, what is the equation that would be used to calculate the price(P) of water on January 1st, 2007?

A. $5P = 2$

B. $\frac{P}{2} = 2$

C. $2p = 2$

D. $2P = 5$

44) The line $2y - 1 = 4x + 5$ and $4y - 1 = 2x + 5$ are?

A. Parallel

B. Perpendicular

C. The same line

D. Neither parallel nor perpendicular

45) A rectangular box measures $9\frac{1}{2}$ feet by $7\frac{1}{5}$ feet. It is divided into four equal parts. What is the area of one of those parts?

A. 4.28

B. 15.80

C. 17.10

D. 68.40

46) Find the slope of the line.

A. $\frac{1}{2}$

B. $\frac{1}{3}$

C. 2

D. $-\frac{1}{2}$

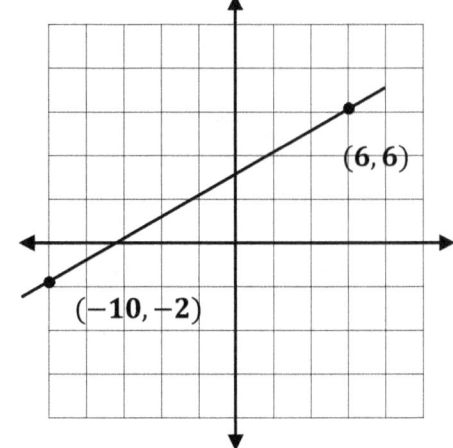

47) Amelia cuts a piece of birthday cake as shown below. What is the volume of the piece of cake?

A. 540 cm^3

B. 270 cm^3

C. 180 cm^3

D. 2,430 cm^3

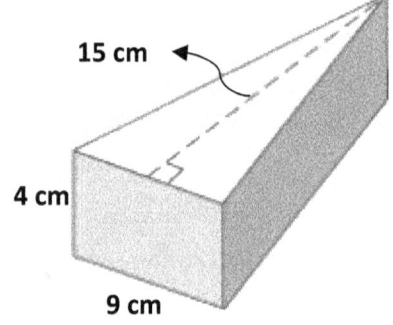

48) Let $f(x) = 5x - 2$. If $f(a) = -12$ and $f(b) = 13$, then what is $f(a + b)$?

A. -2

B. 2

C. 3

D. -3

49) Given that $8x = 6y$, find the ratio $x : y$.

   A. $6 : 48$

   B. $8 : 48$

   C. $4 : 3$

   D. $3 : 4$

50) What is the number of sides of a regular polygon whose interior angles are 144° each? (Remember, the sum of exterior angles of any polygon is 360°).

   A. 5

   B. 10

   C. 6

   D. 12

"End of THEA Practice Test 6."

Answers and Explanations

Answer Key

Now, it's time to review your results to see where you went wrong and what areas you need to improve!

THEA Math Practice Test

Practice Test 1

1	C	20	C	39	A
2	D	21	B	40	D
3	B	22	C	41	B
4	A	23	C	42	D
5	A	24	A	43	A
6	B	25	D	44	D
7	C	26	B	45	C
8	B	27	A	46	B
9	B	28	D	47	A
10	B	29	A	48	C
11	C	30	D	49	C
12	A	31	C	50	A
13	C	32	B		
14	D	33	D		
15	A	34	B		
16	B	35	C		
17	B	36	D		
18	C	37	C		
19	D	38	B		

Practice Test 2

1	C	20	C	39	B
2	B	21	A	40	D
3	D	22	D	41	B
4	A	23	C	42	D
5	C	24	D	43	C
6	D	25	A	44	D
7	A	26	C	45	C
8	A	27	D	46	A
9	C	28	B	47	B
10	B	29	A	48	C
11	D	30	D	49	D
12	C	31	A	50	B
13	C	32	C		
14	D	33	C		
15	A	34	B		
16	C	35	D		
17	D	36	D		
18	C	37	C		
19	B	38	A		

THEA Math Practice Test

Practice Test 3

1	C	20	C	39	A
2	D	21	B	40	D
3	B	22	C	41	B
4	A	23	C	42	D
5	A	24	A	43	A
6	B	25	D	44	D
7	C	26	B	45	C
8	B	27	A	46	B
9	B	28	D	47	A
10	B	29	A	48	C
11	C	30	D	49	C
12	A	31	C	50	A
13	C	32	B		
14	D	33	D		
15	A	34	B		
16	B	35	C		
17	B	36	D		
18	C	37	C		
19	D	38	B		

Practice Test 4

1	C	20	C	39	B
2	B	21	A	40	D
3	D	22	D	41	B
4	A	23	C	42	D
5	C	24	D	43	C
6	D	25	A	44	D
7	A	26	C	45	C
8	A	27	D	46	A
9	C	28	B	47	B
10	B	29	A	48	C
11	D	30	D	49	D
12	C	31	A	50	B
13	C	32	C		
14	D	33	C		
15	A	34	B		
16	C	35	D		
17	D	36	D		
18	C	37	C		
19	B	38	A		

THEA Math Practice Test

Practice Test 5

#	Ans	#	Ans	#	Ans
1	C	20	C	39	A
2	D	21	B	40	D
3	B	22	C	41	B
4	B	23	C	42	D
5	A	24	A	43	A
6	B	25	D	44	D
7	C	26	B	45	C
8	B	27	A	46	B
9	B	28	D	47	A
10	B	29	A	48	C
11	C	30	D	49	C
12	A	31	C	50	A
13	C	32	B		
14	D	33	D		
15	A	34	B		
16	B	35	C		
17	B	36	D		
18	C	37	C		
19	D	38	B		

Practice Test 6

#	Ans	#	Ans	#	Ans
1	C	20	C	39	B
2	B	21	A	40	D
3	D	22	D	41	B
4	A	23	C	42	D
5	C	24	D	43	C
6	D	25	A	44	D
7	A	26	C	45	C
8	A	27	D	46	A
9	C	28	B	47	B
10	B	29	A	48	C
11	D	30	D	49	D
12	C	31	A	50	B
13	C	32	C		
14	D	33	C		
15	A	34	B		
16	C	35	D		
17	D	36	D		
18	C	37	C		
19	B	38	A		

THEA Practice Test 1
Answers and Explanations

1) Answer: C

$\frac{-46}{5} = -9.2$ and $\frac{57}{7} = 8.1$, then the odd numbers are:

$(-9, -7, -5, -3, -1, 1, 3, 5, 7)$

2) Answer: D

The distance between two points always is positive. Use formula:

AB = |b − a| or |a − b| → |−3 − 10| or |10 − (−3)| = |10 + 3|

3) Answer: B

Sum of the measures of the angles of a triangle is 180, if two angles are 60 then the third one is 60, then the triangle is equilateral triangle, and all side are equal.

4) Answer: A

All factors of 12 are: 1, 2, 3, 4, 6, 12, then sum of them is 28.

5) Answer: A

Rewriting each fraction with common denominator or converting each fraction to decimal and order the decimal from least to greatest.

$\frac{5}{9} = 0.56$ $\frac{2}{5} = 0.4$ $\frac{1}{8} = 0.125$ $\frac{25}{28} = 0.89$ $\frac{12}{17} = 0.70$

6) Answer: B

calculating Elena's total earnings: 36 hours × $6.50 an hour = $234

Next, divide this total by her brother's hourly rate:

$234 ÷ $9.75 = 24 hours

7) Answer: C

number of fiction books: x ; number of nonfiction books: $x + 1,500$

Total number of books: $x + (x + 1,500)$

20% of the total number of books are fiction, therefore:

$20\%[x + (x + 1,500)] = x \to 0.2(2x + 1,500) = x$

$0.4x + 300 = x \rightarrow 300 = x - 0.4x \rightarrow 0.6x = 300$

$\rightarrow x = 500$ number of fictions

$x + 1,500 = 500 + 1,500 = 2,000$, the number of nonfiction books

$500 + 2,000 = 2,500$, the total number of books in the library

8) Answer: B

The scale is: 6 cm:1km, (6 cm on the map represents an actual distance of 1 km).

first necessary to rewrite the scale ratio in terms of units squared:

6^2 cm square:1^2 km square, which gives 36 cm^2 : 1 km^2

Then, $\frac{36\ cm^2}{1 km^2} = \frac{90\ cm^2}{x\ km^2}$, (where x is the unknown actual area).

Every proportion you write should maintain consistency in the ratios described (km^2 both occupy the denominator).

Cross-multiply and isolate to solve for the unknown area x:

$x \cdot \frac{36\ cm^2}{1 km^2} = 90\ cm^2 \rightarrow x = 90\ cm^2 \cdot \frac{1\ km^2}{36\ cm^2} \rightarrow x = 2.5\ km^2$

9) Answer: B

move decimal point in divisor so last digit is in the unit place (0.3 to 3)

move decimal point in dividend same number of places to the right,

(15.15 to 151.5); divide (1,515 ÷ 3)=505

insert a decimal point into the answer above the decimal point in the dividend (50.5)

10) Answer: B

Circumference= $2\pi r = 2 \times \pi \times 9 = 18\pi$

11) Answer: C

Be careful with the conversion factor (per hundred gallons; NOT per gallon).

$31,500 \times \frac{0.86}{100} = 270.9$

$270.9 + 5.30 = 276.2$

THEA Practice Tests

12) Answer: A

$\frac{(3x^4-6x^3)}{(x^3-2x^2)} = \frac{3x^3(x-2)}{x^2(x-2)} = 3x$

13) Answer: C

For odd index we can have negative radicand.

In the even index, negative radicand is undefined.

$\sqrt{-25}$ has a negative number under the even index, so it is non-real.

Negative numbers don't have real square roots, because negative and positive integer squared is either positive or 0.

14) Answer: D

$4f(3a) = 864 \rightarrow$ (divide by 4): $f(3a) = 216$

(subtitute 3a) $6(3a)^2 = 216 \rightarrow$ (divide by 6): $(3a)^2 = 36 \rightarrow 9a^2 = 36 \rightarrow$

$a^2 = 4 \rightarrow a = 2$

15) Answer: A

3 hours and 25 minutes is 3.25 hour.

R × T = D → 140 × 3.25 = D → D = 455 miles

16) Answer: B

$5x - 8 \leq 12x + 6 \rightarrow Add\ 8: 5x - 8 + 8 \leq 12x + 6 + 8 \rightarrow$

$subtract\ 12x: 5x - 12x \leq 12x - 12x + 14 \rightarrow -7x \leq 14 \rightarrow x \geq -2$

17) Answer: B

Use formula to raise a number: $(x^a)^b = x^{ab} \rightarrow 2^8 = (2^2)^4 = 4^4$

18) Answer: C

Simple interest rate: I = prt (I = interest, p = principal, r = rate, t = time)

$I = 4,200 \times 0.038 \times 5 = 798$

19) Answer: D

A straight angle is an angle measured exactly 180°

$61° + 42° = 103°; 180° - 103° = 77°$

WWW.MathNotion.com

THEA Practice Tests

20) Answer: C

$$\frac{15x^7y^4z^{-4}}{6x^3y^8z^2} = \frac{15}{6} \times \frac{x^7}{x^3} \times \frac{y^4}{y^8} \times \frac{z^{-4}}{z^2} = \frac{5}{2} \times x^4 \times \frac{1}{y^4} \times \frac{1}{z^6} = \frac{5x^4}{2y^4z^6}$$

21) Answer: B

$$m = \frac{y_2 - y_1}{x_2 - x_1} = \frac{3 - (-2)}{4 - 1} = \frac{5}{3}$$

$$y - y_1 = m(x - x_1) \to y - (-3) = \frac{5}{3}(x - 4)$$

$$y + 3 = \frac{5}{3}(x - 4) \to 3(y + 3) = 5(x - 4) \to 3y + 9 = 5x - 20$$

$$3y - 5x = -20 - 9 \to 3y - 5x = -29$$

22) Answer: C

$$\frac{x^2 + 12}{y - 2} = \frac{(-3)^2 + 12}{5 - 2} = \frac{21}{3} = 7$$

23) Answer: C

Ethan paid $4.2 for three hours and $1.25 for each of three half-hour period after that.

$3 \times 1.25 = 3.75 \to 4.2 + 3.75 = 7.95$

24) Answer: A

Cross multiply and isolate x: $\frac{c-d}{dx} = \frac{3}{5} \to 5(c - d) = 3dx \to x = \frac{5(c-d)}{3d}$

$$x = \frac{5}{3}\left(\frac{c}{d} - \frac{d}{d}\right) = \frac{5}{3}\left(\frac{c}{d} - 1\right)$$

25) Answer: D

$4x^4 - x^2 = x^2(4x^2 - 1) = x^2(2x - 1)(2x + 1)$

26) Answer: B

Factoring: $(x - 7)(x - 9) = 0 \to \begin{cases} x - 7 = 0 \to x = 7 \\ x - 9 = 0 \to x = 9 \end{cases}$

27) Answer: A

Two points are (0,1) and (8,3)

$$m = \frac{y_2 - y_1}{x_2 - x_1} = \frac{3 - 1}{8 - 0} = \frac{2}{8} = \frac{1}{4}$$

$y - y_1 = m(x - x_1) \rightarrow y - 1 = \frac{1}{4}(x - 0)$

$y - 1 = \frac{1}{4}x \rightarrow y = \frac{1}{4}x + 1$

28) Answer: D

$\begin{cases} 2 \times (-4x + 5y = -4) \\ 5 \times (3x - 2y = 3) \end{cases} \rightarrow \begin{cases} -8x + 10y = -8 \\ 15x - 10y = 15 \end{cases}$ →add two equations:

$7x = 7 \rightarrow x = 1$

29) Answer: A

Triangle third side rule: length of the one side of a triangle is less than the sum of the lengths of the other two sides and greater than the positive difference of the lengths of the other two sides.

the third side is less than 7+4=11

and greater than 7−4=3

30) Answer: D

Mean = (121 + 129 + 136 + 149 + 167 + 174) ÷ 6 = 876 ÷ 6 = 146

Only 151 can increase the mean.

31) Answer: C

The mode is the value which occurs with the greatest frequency. Oregon and Ohio are greatest and the same frequency (3 times).

32) Answer: B

convert the given distance, 9 km, into centimeters, (units on the map)

9 km = 9,000 m = 900,000 cm

divide by the ratio 1:60,000.

$\frac{900,000}{60,000} = 15$ cm

33) Answer: D

Area of square is: $a^2 = (\sqrt{11})^2 = 11$

THEA Practice Tests

34) Answer: B

$3x^3y^2 (2x^2y^2)^5 = 3x^3y^2(32x^{10}y^{10}) = 96x^{13}y^{12}$

35) Answer: C

Change percent to decimal: $15\% = 0.15$

$0.15 \times 180 = 27$

36) Answer: D

Using the formula for mean: Mean $= \frac{sum\ of\ the\ several\ given\ values}{number\ of\ value\ given}$

$= \frac{x+(x-6)+(x-6)}{3} = \frac{3x-12}{3} = \frac{3(x-4)}{3} = x - 4$

37) Answer: C

a value that is inversely proportional to another value: $a = \frac{k}{3b-8}$ (where k is a constant of proportionality)

substitute a and b: $12 = \frac{k}{3(6)-8} \to k = 12(10) = 120 \to a = \frac{120}{3b-8}$

38) Answer: B

Baseballs and basketballs are spherical. The volume of sphere: $v = \frac{4}{3}\pi r^3$

$972\pi = \frac{4}{3}\pi r^3 \to 972 = \frac{4}{3}r^3 \to 2,916 = 4r^3 \to r^3 = 729 \to r = 9$

$d = 2r \to d = 2 \times 9 = 18$

39) Answer: A

Use percent formula: Part $= \frac{percent \times whole}{100}$

$168 = \frac{percent \times 120}{100} \Rightarrow \frac{168}{1} = \frac{percent \times 120}{100}$, cross multiply.

$16,800 = $ percent $\times 120$, divide both sides by 120. $\to$ percent$= 140$

40) Answer: D

The sample shows that 4 out of 300 phones will be faulty. Consequently, a proportion can be set up.

$\frac{4}{300} = \frac{P}{15,000}$ (P: Faulty Phone) $\to P = \frac{4 \times 15,000}{300} = 200$

THEA Practice Tests

41) Answer: B

We can Write an equation to solve the problem.

Emma Books = Mia books +3 → Mia book = Emma − 3 = b − 3

Emma + Mia = 25

$b + b - 3 = 25 \rightarrow 2b - 3 = 25 \rightarrow 2b = 25 + 3 \rightarrow b = 14$

42) Answer: D

Probability = $\frac{number\ of\ desired\ outcomes}{number\ of\ total\ outcomes} = \frac{7}{4+3+7} = \frac{7}{14} = \frac{1}{2} = 0.5$

43) Answer: A

Surface Area of a cylinder = $2\pi r(r+h)$,

The radius of the cylinder is 3 inches, and its height are 7 inches. π is 3.14. Then:

Surface Area of a cylinder = 2 (3.14) (3) (3 + 7) = 188.4 inches

44) Answer: D

A line perpendicular to a line with slope m has a slope of $-\frac{1}{m}$.

So, the slope of the line perpendicular to the given line is $-\frac{1}{\frac{a}{b}} = -\frac{b}{a}$.

45) Answer: C

If 22% of the total number of customers is female, then 100% − 22% = 78% of the customers are male. Calculate 78% of the total. 0.78 × 800 = 624.

46) Answer: B

First, add the numbers given in the proportion to get a denominator: 7 + 4 = 11.

Then, the two parts can be represented as $\frac{7}{11}$ and $\frac{4}{11}$.

The smaller share is $\frac{4}{11}$ of 72.3 kg: $\frac{4}{11} \times 72.3\ kg = 26.29$

47) Answer: A

Since this is subtraction of 2 fractions with different denominators, their least common denominator is: $(x-1)(2x+3)$

$\frac{3}{(x-1)} - \frac{4}{(2x+3)} = \frac{3(2x+3)-4(x-1)}{(x-1)(2x+3)} = \frac{6x+9-4x+4}{(x-1)(2x+3)} = \frac{2x+13}{(x-1)(2x+3)}$

48) Answer: C

State the problem in a mathematical equation:

$3x = 7x - 16 \rightarrow 3x - 7x = -16 \rightarrow -4x = -16 \rightarrow x = \dfrac{-16}{-4} = 4$

$7x - 16 = 7(4) - 16 = 12$

49) Answer: C

$4h = 3p + 4$:

$p = 4 \rightarrow 4h = 3(4) + 4 = 16 \rightarrow 4h = 16 \rightarrow h = 4$

$p = 8 \rightarrow 4h = 3(8) + 4 = 28 \rightarrow 4h = 28 \rightarrow h = 7$

$p = 12 \rightarrow 4h = 3(12) + 4 = 40 \rightarrow 4h = 40 \rightarrow h = 10$

possible values of h: $\{4, 7, 10\}$

50) Answer: A

There are 6 digits in the repeating decimal (076923), so 7 would be the second, eighth, fourteenth digit and so on.

To find the 440st digit, divide 440 by 6. $440 \div 6 = 73$ R2

Since the remainder is 2, that means the 440st digit is the same as the 2nd digit, which is 7.

THEA Practice Test 2
Answers and Explanations

1) Answer: C

The smallest prime number is 2, and the largest even negative integer is −2.

2+6 (−2) = 2−12 = −10.

2) Answer: B

The difference between his income and his cost is monthly saving.

January: $2,780− $1,002= $1,778

February: $2,895 − $975= $1,920

March: $2,990 − $1,012 = $1,978

3) Answer: D

State the problem in a mathematical sentence:

$a + 38 = 272 - 5a$

$a + 5a = 272 - 38$

$6a = 234 \to a = 39$

4) Answer: A

$7\frac{5}{32} - 4\frac{1}{4} + 3\frac{3}{8} = (7 - 4 + 3)\frac{5}{32} - \frac{8}{32} + \frac{12}{32} = 6(\frac{-3}{32} + \frac{12}{32}) = 6(\frac{12-3}{32}) = 6\frac{9}{32}$

5) Answer: C

$5^2 + 5 + 5^0 = 25 + 5 + 1 = 31$

6) Answer: D

$\frac{133}{190} = \frac{133}{19} \times \frac{1}{10} = 7 \times \frac{1}{10} = 0.7$

7) Answer: A

Use the formula for Percent of Change:

$\frac{New\ Value - Old\ Value}{Old\ Value} \times 100\% = \frac{42.40-40}{40} \times 100\% = \frac{2.40}{40} \times 100\% = 6\%$

THEA Practice Tests

8) Answer: A

$|3x - 6| = 9 \rightarrow \begin{cases} 3x - 6 = 9 \rightarrow 3x = 15 \rightarrow x = 5 \\ 3x - 6 = -9 \rightarrow 3x = -3 \rightarrow x = -1 \end{cases}$

9) Answer: C

Multiply equation (2) by 4. Add two equations [(1) +4(2)]:

$\begin{cases} x - 4y = 6 \\ 12x + 4y = 20 \end{cases} \rightarrow 13x = 26 \rightarrow x = 2$

Substitute $x = 2$ into equation (1): $2 - 4y = 6 \rightarrow -4y = 4 \rightarrow y = -1$

10) Answer: B

4 hours equals 240 minutes. The train passes 3 stations every 20 minutes: $\frac{3}{20}$

$\frac{3}{20} = \frac{x}{240} \rightarrow x = \frac{240 \times 3}{20} = 36$ stations

11) Answer: D

Difference in percent: $12\% - 10\% = 2\%$

$2\% \times 170 = 0.02 \times 170 = 3.40$

12) Answer: C.

$C = \pi d = 2\pi r = 2\pi \times 29 = 58\pi$

13) Answer: C

$V = l \times w \times h = 23 \times 10 \times 13 = 2{,}990 \; in^3$

14) Answer: D

$\sqrt[3]{3^{-9}} = \sqrt[3]{\frac{1}{3^9}} = \frac{\sqrt[3]{1}}{\sqrt[3]{3^9}} = \frac{1}{3^{\left(\frac{9}{3}\right)}} = \frac{1}{3^3} = 3^{-3} = \frac{1}{27}$

15) Answer: A

There are no values of the variable that make the equation true.

16) Answer: C

Volume of cylinder: $V = \pi r^2 h = \pi \times 6^2 \times 10 = 360\pi$

Volume of cone: $V = \frac{1}{3}\pi r^2 h = \frac{1}{3}\pi \times 6^2 \times 4 = 48\pi$

$360\pi + 48\pi = 408\pi$.

17) Answer: D

$7.59 \times 10^{-3} = 0.00759$

18) Answer: C

$x + 49 + 46 + 39 = 180 \rightarrow x + 134 = 180 \rightarrow x = 46$

19) Answer: B

use the Pythagorean theorem to find the value of unknown side.

$a^2 + b^2 = c^2 \rightarrow 30^2 = a^2 + 18^2 \rightarrow a^2 = 900 - 324 = 576 \rightarrow a = 24$

20) Answer: C

Use percent formula: Part $= \frac{\text{percent} \times \text{whole}}{100} \rightarrow$ Part $= \frac{15 \times 160}{100} = 24$

Last price: $160 + 24 = \$184$

21) Answer: A

Area of trapezoid: $\frac{(a+b)}{2} \times h \rightarrow A = \frac{(10+13)}{2} \times 8 = 92$

Area of triangle: $\frac{b \times h}{2} \rightarrow A = \frac{6 \times 8}{2} = 24$

Area of shaded region: $92 - 24 = 68$

22) Answer: D

$xy - 9x = 42 \rightarrow x(y - 9) = 42$

$7x = 42 \rightarrow x = 6$

23) Answer: C

Multiply both sides by 5: $5 \times \left(\frac{x}{5}\right) = 5 \times (x - 8) \rightarrow x = 5x - 40$

Subtract both side by x and add 40 to both sides: $40 + x - x = 5x - x - 40 +$

$40 \rightarrow 4x = 40 \rightarrow x = 10$

24) Answer: D

$\begin{cases} 5 \times (-7x + 3y = 8) \\ -3 \times (-6x + 5y = 2) \end{cases} \rightarrow \begin{cases} -35x + 15y = 40 \\ 18x - 15y = -6 \end{cases} \rightarrow$ add two equations:

$-17x = 34 \rightarrow x = -2$

THEA Practice Tests

25) Answer: A

There are 4 parts labeled "H" out of a total of 8 equal parts.

The probability of not spinning at "H" is 4 out of 8.

26) Answer: C

The best estimate of the product of 42 and 4.06 to the nearest whole number.

Since 42 is already a whole number and 4.06 is closer to 4 than 5. Option C is the correct answer.

27) Answer: D

Fuel consumption rate $= \frac{9}{60} = 0.15$ liter per mile

Cost: $1,400 \times 0.15 \times 1.20 = 252$

28) Answer: B

Point $1(x_A, y_A) = (-3, -5)$; Point $2(x_B, y_B) = (1, -2)$

Distance between two points $= \sqrt{(x_B - x_A)^2 + (y_B - y_A)^2}$

$\to d = \sqrt{(1-(-3))^2 + (-2-(-5))^2} = \sqrt{4^2 + 3^2} = \sqrt{16 + 9}$

$\to d = \sqrt{25} = 5$

29) Answer: A

Ordered values: 98, 156, 188, 215, 230, 389, 401.

The 4th number is median.

30) Answer: D

$\sqrt{5x - 6} = 7 \to 5x - 6 = 49 \to 5x = 55 \to x = 11$

31) Answer: A

Get the equation into slop intercept form:

$y = mx + b$ where m is slope and b is y-intercept.

$x = 3y - 9$; Add 9 to both sides $\to x + 9 = 3y$; Dividing by 3: $\to y = \frac{1}{3}x + 3$

Thus, the y-intercept is 3.

THEA Practice Tests

32) Answer: C

$2^{3x} = 64 \to (2^3)^x = 64 \to 8^x = 8^2 \to x = 2$

33) Answer: C

$32: 2 \times 2 \times 2 \times 2 \times 2$

$48: 2 \times 2 \times 2 \times 2 \times 3$

LCM $(48, 32) = 2 \times 2 \times 2 \times 2 \times 2 \times 3 = 96$

34) Answer: B

Surface area $(SA): 6a^2 \to 150 = 6a^2 \to a^2 = 25 \to a = 5$ cm

Volume $(v): a^3 \to v = 5^3 = 125$ cm³

35) Answer: D

Write the equation into standard form: $x^2 + 4x = 32 \to x^2 + 4x - 32 = 0$

Factor this expression: $(x-4)(x+8) = 0 \to x = 4 \text{ or } x = -8 \to x^2 = \begin{cases} 16 \\ 64 \end{cases}$

36) Answer: D

The minimum amount of water: $8 \times \frac{7}{8} = 7; 7 \div 4 = 1.75$

The maximum amount of water: $8 \times \frac{7}{8} = 7$

Amount of water in all pitchers: $1.75 < w < 7$

37) Answer: C

The range is: highest value – lowest value.

R = 51 − 16 = 35

38) Answer: A

$d = 2r \to r = \frac{d}{2} = \frac{22}{2} = 11$

Area of circle $(A): \pi r^2 = \pi \times 11^2 = 379.94 \cong 379.9$

39) Answer: B

Factor the expression: $\frac{3x^2 - 14x - 5}{9(x^2 - \frac{1}{9})} = \frac{(x-5)(3x+1)}{9x^2 - 1} = \frac{(x-5)(3x+1)}{(3x-1)(3x+1)} = \frac{x-5}{3x-1}$

THEA Practice Tests

40) Answer: D

$(\frac{1}{2})^{-5} = \frac{1}{2^{-5}} = 2^5 = 32$

41) Answer: B

$\frac{(6x^{-4}y^2)^2}{252y^{-3}z^0} = \frac{36x^{-8}y^4}{252y^{-3}z^0} = \frac{y^7}{7x^8}$

42) Answer: D

Rate: $\frac{\$1}{1Kg} = \frac{100¢}{1,000g} \rightarrow 1\ \$/kg = 0.1\ ¢/g$

The price is: $20 \times 0.1 = 2$ cent per kilogram.

43) Answer: C

$4P = 8 \rightarrow P = 2$ The price of 2007 is one-fourth of 2012 ($8 \div 4 = 2$)

44) Answer: D

First equation: $3y + 4 = 18x + 7 \rightarrow 3y = 18x + 3 \rightarrow y = 6x + 1 \rightarrow m_1 = 6$

Second question: $6y - 2 = x + 3 \rightarrow 6y = x + 5 \rightarrow y = \frac{1}{6}x + \frac{5}{6} \rightarrow m_2 = \frac{1}{6}$

$m_1 = 6$ and $m_2 = \frac{1}{6}$, they aren't equal slopes or negative reciprocals.

45) Answer: C

Area $= 7\frac{1}{3} \times 6\frac{3}{4} = \frac{22}{3} \times \frac{27}{4} = \frac{594}{12} = \frac{99}{2}$ square feet.

One-fourth $= \frac{99}{2} \div 3 = \frac{99}{2} \times \frac{1}{3} = \frac{99}{6} = \frac{33}{2} = 16.5$

46) Answer: A

Two points are $(6, 2)$ and $(-6, -4) \rightarrow m = \frac{y_2 - y_1}{x_2 - x_1} = \frac{-4 - 2}{-6 - 6} = \frac{-6}{-12} = \frac{1}{2}$

47) Answer: B

$V = \left(\frac{base \times height}{2}\right) \times$ height of prism $\rightarrow V = \left(\frac{8 \times 17}{2}\right) \times 6 = 408$

48) Answer: C

$f(a) = 3a - 6$ and $f(a) = -15$: $3a - 6 = -15$ so that $3a = -9$ and $a = -3$.

$f(b) = 3b - 6$ and $f(b) = 6$: $3b - 6 = 6$ so that $3b = 12$ and $b = 4$

THEA Practice Tests

Finally, f(a + b) = f(−3 + 4) = f(1)

f(1) = 3(1) − 6 = −3.

49) Answer: D

The ratio of $x:y$ is equivalent to x divided by y, or $\frac{x}{y}$.

Equation: $10x = 4y$

Dividing both sides by 10: $x = \frac{4y}{10}$

Dividing both sides by y: $\frac{x}{y} = \frac{4}{10} = \frac{2}{5}$

Then, $x:y$ is $2:5$

50) Answer: B

The formula for measurement of each angle of a regular polygon:

$x = \frac{180(n-2)}{n}$, (x is the measurement of the interior angle; n is the number of sides)

Substituting the given information:

$\frac{180(n-2)}{n} = 162 \rightarrow 180n - 360 = 162n \rightarrow 180n - 162n = 360$

$\rightarrow 18n = 360 \rightarrow n = 20$

The polygon has 20 angles and 20 sides.

THEA Practice Tests

THEA Practice Test 3
Answers and Explanations

1) Answer: C

$\frac{-69}{6} = -11.5$ and $\frac{19}{5} = 3.8$, then the odd numbers are:

$(-11, -9, -7, -5, -3, -1, 1, 3)$

2) Answer: D

The distance between two points always is positive. Use formula:

AB = |b − a| or |a − b| → |− 4 − 14| or |14 − (−4)| = |14 + 4|

3) Answer: B

Sum of the measures of the angles of a triangle is 180, if two angles are 60 then the third one is 60, then the triangle is equilateral triangle, and all side are equal.

4) Answer: A

All factors of 21 are: 1, 3, 7, 21, then sum of them is 32.

5) Answer: A

Rewriting each fraction with common denominator or converting each fraction to decimal and order the decimal from least to greatest.

$\frac{3}{8} = 0.375$ $\frac{4}{7} = 0.57$ $\frac{1}{5} = 0.2$ $\frac{23}{25} = 0.92$ $\frac{14}{19} = 0.73$

6) Answer: B

calculating Elena's total earnings: 33 hours × $7.00 an hour = $231

Next, divide this total by her brother's hourly rate:

$231 ÷ $8.25 = 28 hours

7) Answer: C

number of fiction books: x ; number of nonfiction books: $x + 1,200$

Total number of books: $x + (x + 1,200)$

30% of the total number of books are fiction, therefore:

$30\%[x + (x + 1,200)] = x \rightarrow 0.3(2x + 1,200) = x$

WWW.MathNotion.com

THEA Practice Tests

$0.6x + 360 = x \rightarrow 360 = x - 0.6x \rightarrow 0.4x = 360$

$\rightarrow x = 900$ number of fictions

$x + 1,200 = 900 + 1,200 = 2,100$, the number of nonfiction books

$900 + 2,100 = 3,000$, the total number of books in the library

8) Answer: B

The scale is: 5 cm:1km, (5 cm on the map represents an actual distance of 1 km).

first necessary to rewrite the scale ratio in terms of units squared:

5^2cm square:1^2km square, which gives $25\ cm^2 : 1\ km^2$

Then, $\frac{25\ cm^2}{1km^2} = \frac{85\ cm^2}{x\ km^2}$, (where x is the unknown actual area).

Every proportion you write should maintain consistency in the ratios described (km^2 both occupy the denominator).

Cross-multiply and isolate to solve for the unknown area x:

$x \cdot \frac{25\ cm^2}{1km^2} = 85\ cm^2 \rightarrow x = 85\ cm^2 \cdot \frac{1\ km^2}{25\ cm^2} \rightarrow x = 3.4\ km^2$

9) Answer: B

move decimal point in divisor so last digit is in the unit place (0.9 to 9)

move decimal point in dividend same number of places to the right,

(45.45 to 454.5); divide (4,545 ÷ 9)=505

insert a decimal point into the answer above the decimal point in the dividend (50.5)

10) Answer: B

Circumference= $2\pi r = 2 \times \pi \times 11 = 22\pi$

11) Answer: C

Be careful with the conversion factor (per hundred gallons; NOT per gallon).

$28,500 \times \frac{0.72}{100} = 205.2$

$205.2 + 4.40 = 209.6$

THEA Practice Tests

12) Answer: A

$\frac{(2x^5+6x^4)}{(x^4+3x^2)} = \frac{2x^4(x+3)}{x^3(x+3)} = 2x$

13) Answer: C

For odd index we can have negative radicand.

In the even index, negative radicand is undefined.

$\sqrt{-49}$ has a negative number under the even index, so it is non-real.

Negative numbers don't have real square roots, because negative and positive integer squared is either positive or 0.

14) Answer: D

$5f(2a) = 180 \to$ (divide by 5): $f(2a) = 36$

(subtitute 2a) $3(2a)^2 = 108 \to$ (divide by 3): $(2a)^2 = 36 \to 4a^2 = 36 \to$

$a^2 = 9 \to a = 3$

15) Answer: A

4 hours and 45 minutes is 4.45 hour.

$R \times T = D \to 160 \times 4.45 = D \to D = 712$ miles

16) Answer: B

$9x - 7 \leq 15x + 11 \to$ Add 7: $9x - 7 + 7 \leq 15x + 7 + 11 \to$

subtract $15x$: $9x - 15x \leq 15x - 15x + 18 \to -6x \leq 18 \to x \geq -3$

17) Answer: B

Use formula to raise a number: $(x^a)^b = x^{ab} \to 3^6 = (3^2)^3 = 9^3$

18) Answer: C

Simple interest rate: I = prt (I = interest, p = principal, r = rate, t = time)

$I = 3{,}600 \times 0.045 \times 2 = 324$

19) Answer: D

A straight angle is an angle measured exactly 180°

$72° + 34° = 106° \to 180° - 106° = 74°$

WWW.MathNotion.com

THEA Practice Tests

20) Answer: C

$$\frac{21x^9 y^2 z^{-3}}{14\, x^5 y^6 z^3} = \frac{21}{14} \times \frac{x^9}{x^5} \times \frac{y^4}{y^8} \times \frac{z^{-3}}{z^3} = \frac{3}{2} \times x^4 \times \frac{1}{y^4} \times \frac{1}{z^6} = \frac{3x^4}{2\, y^4 z^6}$$

21) Answer: B

$$m = \frac{y_2 - y_1}{x_2 - x_1} = \frac{1-(-4)}{5-2} = \frac{5}{3}$$

$$y - y_1 = m(x - x_1) \to y - (-4) = \frac{5}{3}(x-2)$$

$$y + 4 = \frac{5}{3}(x-2) \to 3(y+4) = 5(x-2) \to 3y + 12 = 5x - 10$$

$$3y - 5x = -10 - 12 \to 3y - 5x = -22$$

22) Answer: C

$$\frac{x^2 + 7}{y - 3} = \frac{(-5)^2 + 7}{7 - 3} = \frac{32}{4} = 8$$

23) Answer: C

Ethan paid $3.2 for two hours and $1.5 for each of three half-hour period after that.

$$3 \times 1.5 = 4.5 \to 3.2 + 4.5 = 7.7$$

24) Answer: A

Cross multiply and isolate x: $\frac{c-d}{dx} = \frac{2}{7} \to 7(c-d) = 2dx \to x = \frac{7(c-d)}{2d}$

$$x = \frac{7}{2}\left(\frac{c}{d} - \frac{d}{d}\right) = \frac{7}{2}\left(\frac{c}{d} - 1\right)$$

25) Answer: D

$$9x^4 - 4x^2 = x^2(9x^2 - 4) = x^2(3x-2)(3x+2)$$

26) Answer: B

Factoring: $(x-6)(x-8) = 0 \to \begin{cases} x - 6 = 0 \to x = 6 \\ x - 8 = 0 \to x = 8 \end{cases}$

27) Answer: A

Two points are (0,4) and (10,6)

$$m = \frac{y_2 - y_1}{x_2 - x_1} = \frac{6-4}{10-0} = \frac{2}{10} = \frac{1}{5}$$

THEA Practice Tests

$$y - y_1 = m(x - x_1) \rightarrow y - 4 = \frac{1}{5}(x - 0)$$

$$y - 4 = \frac{1}{5}x \rightarrow y = \frac{1}{5}x + 4$$

28) Answer: D

$$\begin{cases} 3 \times (-3x + 2y = -2) \\ 2 \times (4x - 3y = 5) \end{cases} \rightarrow \begin{cases} -9x + 6y = -6 \\ 8x - 6y = 10 \end{cases} \rightarrow \text{add two equations:}$$

$$-x = 4 \rightarrow x = -4$$

29) Answer: A

Triangle third side rule: length of the one side of a triangle is less than the sum of the lengths of the other two sides and greater than the positive difference of the lengths of the other two sides.

the third side is less than 5+8=13 and greater than 8−5=3

30) Answer: D

Mean = $(111 + 120 + 127 + 140 + 159 + 165) \div 6 = 822 \div 6 = 137$

Only 148 can increase the mean.

31) Answer: C

The mode is the value which occurs with the greatest frequency. Oregon and New York are greatest and the same frequency (3 times).

32) Answer: B

convert the given distance, 8 km, into centimeters, (units on the map)

8 km = 8,000 m = 800,000 cm

divide by the ratio 1:50,000.

$$\frac{800,000}{50,000} = 16\text{cm}$$

33) Answer: D

Area of square is: $a^2 = (\sqrt{13})^2 = 13$

34) Answer: B

$2x^4y^4 (3x^3y^3)^4 = 2x^4y^4(81x^{12}y^{12}) = 162x^{16}y^{16}$

THEA Practice Tests

35) Answer: C

Change percent to decimal: $25\% = 0.25$;

$0.25 \times 220 = 55$

36) Answer: D

Using the formula for mean: Mean $= \frac{sum\ of\ the\ several\ given\ values}{number\ of\ value\ given}$

$= \frac{x+(x-3)+(x-3)}{3} = \frac{3x-6}{3} = \frac{3(x-2)}{3} = x - 2$

37) Answer: C

a value that is inversely proportional to another value:

$a = \frac{k}{2b-5}$ (Where k is a constant of proportionality)

substitute a and b: $8 = \frac{k}{2(9)-5} \rightarrow k = 8(13) = 104 \rightarrow a = \frac{104}{2b-5}$

38) Answer: B

Baseballs and basketballs are spherical. The volume of sphere: $v = \frac{4}{3}\pi r^3$

$36\pi = \frac{4}{3}\pi r^3 \rightarrow 36 = \frac{4}{3}r^3 \rightarrow 108 = 4r^3 \rightarrow r^3 = 27 \rightarrow r = 3$

$d = 2r \rightarrow d = 2 \times 3 = 6$

39) Answer: A

Use percent formula: Part $= \frac{percent \times whole}{100}$

$154 = \frac{percent \times 140}{100} \Rightarrow \frac{154}{1} = \frac{percent \times 140}{100}$, cross multiply.

$15,400 = percent \times 140$,

divide both sides by 140. $\rightarrow$ percent $= 110$

40) Answer: D

The sample shows that 5 out of 600 phones will be faulty. Consequently, a proportion can be set up.

$\frac{5}{600} = \frac{P}{18,000}$ (P: Faulty Phone) $\rightarrow P = \frac{5 \times 18,000}{600} = 150$

THEA Practice Tests

41) Answer: B

We can Write an equation to solve the problem.

Emma Books = Mia books + 4 → Mia book = Emma − 4 = b − 4

Emma + Mia = 28

b + b − 4 = 25 → 2b − 4 = 28 → 2b = 28 + 4 → b = 16

42) Answer: D

Probability = $\frac{number\ of\ desired\ outcomes}{number\ of\ total\ outcomes} = \frac{6}{11+6+7} = \frac{6}{24} = \frac{1}{4} = 0.25$

43) Answer: A

Surface Area of a cylinder = 2πr (r + h),

The radius of the cylinder is 4 inches, and its height are 9 inches. π is 3.14. Then:

Surface Area of a cylinder = 2 (3.14) (4) (4 + 9) = 326.56 inches

44) Answer: D

A line perpendicular to a line with slope m has a slope of $-\frac{1}{m}$.

So, the slope of the line perpendicular to the given line is $-\frac{1}{\frac{a}{b}} = -\frac{b}{a}$.

45) Answer: C

If 38% of the total number of customers is female, then 100% − 38% = 62% of the customers are male. Calculate 62% of the total. 0.62 × 750 = 465.

46) Answer: B

First, add the numbers given in the proportion to get a denominator: 8 + 3 = 11.

Then, the two parts can be represented as $\frac{8}{11}$ and $\frac{3}{11}$.

The smaller share is $\frac{3}{11}$ of 66.33 kg: $\frac{3}{11} \times 66.33\ kg = 18.09$

47) Answer: A

Since this is subtraction of 2 fractions with different denominators, their least common denominator is: $(x + 2)(4x − 1)$

$\frac{5}{(x+2)} - \frac{2}{(4x-1)} = \frac{5(4x-1)-2(x+2)}{(x+2)(4x-1)} = \frac{20x-5-2x-4}{(x+2)(4x-1)} = \frac{18x-9}{(x+2)(4x-1)}$

WWW.MathNotion.com

THEA Practice Tests

48) Answer: C

State the problem in a mathematical equation:

$2x = 6x - 20 \rightarrow 2x - 6x = -20 \rightarrow -4x = -20 \rightarrow x = \dfrac{-20}{-4} = 5$

$6x - 20 = 6(5) - 20 = 10$

49) Answer: C

$2h = 5p + 1$:

$p = 3 \rightarrow 2h = 5(3) + 1 = 16 \rightarrow 2h = 16 \rightarrow h = 8$

$p = 5 \rightarrow 2h = 5(5) + 1 = 26 \rightarrow 2h = 26 \rightarrow h = 13$

$p = 11 \rightarrow 2h = 5(11) + 1 = 56 \rightarrow 2h = 56 \rightarrow h = 28$

possible values of h: $\{8, 13, 28\}$

50) Answer: A

There are 6 digits in the repeating decimal (047619), so 4 would be the second, eighth, fourteenth digit and so on.

To find the 764th digit, divide 764 by 6.

764 ÷ 6 = 127 R2

Since the remainder is 2, that means the 764th digit is the same as the 2nd digit, which is 4.

THEA Practice Test 4
Answers and Explanations

1) Answer: C

The smallest prime number is 2, and the largest even negative integer is −2.

2+5 (−2) = 2−10 = −8.

2) Answer: B

The difference between his income and his cost is monthly saving.

January: $3,740− $2,202= $1,538

February: $3,855 − $1,985= $1,870

March: $3,970 − $2,005 = $1,965

3) Answer: D

State the problem in a mathematical sentence:

$a + 29 = 244 - 4a \rightarrow a + 4a = 244 - 29$

$5a = 215 \rightarrow a = 43$

4) Answer: A

$8\frac{7}{24} - 6\frac{2}{3} + 2\frac{5}{6} = (8 - 6 + 2)\frac{7}{24} - \frac{16}{24} + \frac{20}{24} = 4(\frac{-9}{24} + \frac{20}{24}) = 4(\frac{20-9}{24}) = 4\frac{11}{24}$

5) Answer: C

$3^3 + 3 + 3^0 = 27 + 3 + 1 = 31$

6) Answer: D

$\frac{153}{170} = \frac{153}{17} \times \frac{1}{10} = 9 \times \frac{1}{10} = 0.9$

7) Answer: A

Use the formula for Percent of Change:

$\frac{New\ Value - Old\ Value}{Old\ Value} \times 100\% = \frac{32.10 - 30}{30} \times 100\% = \frac{2.10}{30} \times 100\% = 7\%$

8) Answer: A

$|2x - 3| = 7 \rightarrow \begin{cases} 2x - 3 = 7 \rightarrow 2x = 10 \rightarrow x = 5 \\ 2x - 3 = -7 \rightarrow 2x = -4 \rightarrow x = -2 \end{cases}$

THEA Practice Tests

9) Answer: C

Multiply equation (2) by 5. Add two equations [(1) +5(2)]:

$\begin{cases} 2x - 5y = 16 \\ 10x + 5y = 20 \end{cases} \to 12x = 36 \to x = 3$

Substitute $x = 3$ into equation (1): $2(3) - 5y = 16 \to -5y = 10 \to y = -2$

10) Answer: B

3 hours equals 180 minutes. The train passes 5 stations every 30 minutes: $\frac{5}{30}$

$\frac{5}{30} = \frac{x}{180} \to x = \frac{180 \times 5}{30} = 30$ stations

11) Answer: D

Difference in percent: $16\% - 11\% = 5\%$

$5\% \times 160 = 0.05 \times 160 = 8$

12) Answer: C.

$C = \pi d = 2\pi r = 2\pi \times 18 = 36\pi$

13) Answer: C

$V = l \times w \times h = 19 \times 15 \times 8 = 2,280 \ in^3$

14) Answer: D

$\sqrt[4]{5^{-8}} = \sqrt[4]{\frac{1}{5^8}} = \frac{\sqrt[4]{1}}{\sqrt[4]{5^8}} = \frac{1}{5^{\frac{8}{4}}} = \frac{1}{5^2} = 5^{-2} = \frac{1}{25}$

15) Answer: A

There are no values of the variable that make the equation true.

16) Answer: C

Volume of cylinder: $V = \pi r^2 h = \pi \times 9^2 \times 8 = 648\pi$

Volume of cone: $V = \frac{1}{3}\pi r^2 h = \frac{1}{3}\pi \times 9^2 \times 6 = 162\pi$

$648\pi + 162\pi = 810\pi.$

17) Answer: D

$7.25 \times 10^{-5} = 0.0000725$

18) Answer: C

$x + 52 + 42 + 36 = 180 \rightarrow x + 130 = 180 \rightarrow x = 50$

19) Answer: B

use the Pythagorean theorem to find the value of unknown side.

$a^2 + b^2 = c^2 \rightarrow 25^2 = a^2 + 15^2 \rightarrow a^2 = 625 - 225 = 400 \rightarrow a = 20$

20) Answer: C

Use percent formula: Part $= \frac{\text{percent} \times \text{whole}}{100}$

Part $= \frac{12 \times 150}{100} = 18$

Last price: $150 + 18 = \$168$

21) Answer: A

Area of trapezoid: $\frac{(a+b)}{2} \times h \rightarrow A = \frac{(12+14)}{2} \times 10 = 130$

Area of triangle: $\frac{b \times h}{2} \rightarrow A = \frac{7 \times 10}{2} = 35$

Area of shaded region: $130 - 35 = 95$

22) Answer: D

$xy - 8x = 54 \rightarrow x(y - 8) = 54$

$6x = 54 \rightarrow x = 9$

23) Answer: C

Multiply both sides by 4: $4 \times \left(\frac{x}{4}\right) = 4 \times (x - 6) \rightarrow x = 4x - 24$

Subtract both side by x and add 24 to both sides: $24 + x - x = 4x - x - 24 + 24 \rightarrow 3x = 24 \rightarrow x = 8$

24) Answer: D

$\begin{cases} 3 \times (-5x + 4y = 7) \\ -4 \times (-4x + 3y = 5) \end{cases} \rightarrow \begin{cases} -15x + 12y = 21 \\ 16x - 12y = -20 \end{cases}$ →add two equations:

$x = 21 - 20 \rightarrow x = 1$

THEA Practice Tests

25) Answer: A

There are 3 parts labeled "F" out of a total of 8 equal parts.

The probability of not spinning at "F" is 5 out of 8.

26) Answer: C

The best estimate of the product of 32 and 5.04 to the nearest whole number.

Since 32 is already a whole number and 5.04 is closer to 5 than 6. Option C is the correct answer.

27) Answer: D

Fuel consumption rate $= \frac{7}{50} = 0.14$ liter per mile

Cost: $1{,}500 \times 0.14 \times 1.40 = 294$

28) Answer: B

Point $1(x_A, y_A) = (-6, -9)$; Point $2(x_B, y_B) = (2, -3)$

Distance between two points $= \sqrt{(x_B - x_A)^2 + (y_B - y_A)^2}$

$\rightarrow d = \sqrt{(2-(-6))^2 + (-3-(-9))^2} = \sqrt{8^2 + 6^2} = \sqrt{64 + 36}$

$\rightarrow d = \sqrt{100} = 10$

29) Answer: A

Ordered values: 75, 89, 124, 198, 225, 280, 512.

The 4th number is median.

30) Answer: D

$\sqrt{4x - 7} = 9 \rightarrow 4x - 7 = 81 \rightarrow 4x = 88 \rightarrow x = 22$

31) Answer: A

Get the equation into slop intercept form:

$y = mx + b$ where m is slope and b is y-intercept.

$x = 4y - 8$; Add 8 to both sides $\rightarrow x + 8 = 4y$; Dividing by 4: $\rightarrow y = \frac{1}{4}x + 2$

Thus, the y-intercept is 2.

THEA Practice Tests

32) Answer: C

$3^{2x} = 729 \to (3^2)^x = 729 \to 9^x = 9^3 \to x = 3$

33) Answer: C

$18: 2 \times 3 \times 3$

$64: 2 \times 2 \times 2 \times 2 \times 2 \times 2$

LCM $(18, 64) = 2 \times 2 \times 2 \times 2 \times 2 \times 2 \times 3 \times 3 = 576$

34) Answer: B

Surface area $(SA): 6a^2 \to 96 = 6a^2 \to a^2 = 16 \to a = 4$cm

Volume $(v): a^3 \to v = 4^3 = 64$ cm^3

35) Answer: D

Write the equation into standard form: $x^2 + 2x = 35 \to x^2 + 2x - 35 = 0$

Factor this expression: $(x-5)(x+7) = 0 \to x = 5 \text{ or } x = -7 \to x^2 = \begin{cases} 25 \\ 49 \end{cases}$

36) Answer: D

The minimum amount of water: $5 \times \frac{3}{5} = 3; 3 \div 2 = 1.5$

The maximum amount of water: $5 \times \frac{3}{5} = 3$

Amount of water in all pitchers: $1.5 < w < 3$

37) Answer: C

The range is: highest value – lowest value.

R $= 65 - 17 = 48$

38) Answer: A

$d = 2r \to r = \frac{d}{2} = \frac{16}{2} = 8$

Area of circle $(A): \pi r^2 = \pi \times 8^2 = 200.96 \cong 201$

39) Answer: B

Factor the expression: $\frac{2x^2 - 15x + 7}{4(x^2 - \frac{1}{4})} = \frac{(x-7)(2x-1)}{4x^2 - 1} = \frac{(x-7)(2x-1)}{(2x-1)(2x+1)} = \frac{x-7}{2x+1}$

THEA Practice Tests

40) Answer: D

$(\frac{1}{3})^{-4} = \frac{1}{3^{-4}} = 3^4 = 81$

41) Answer: B

$\frac{(5x^{-3}y^4)^3}{100y^{-2}z^2} = \frac{125x^{-9}y^{12}}{100y^{-2}z^2} = \frac{5y^{14}}{4x^9 z^2}$

42) Answer: D

Rate: $\frac{\$1}{1Kg} = \frac{100¢}{1,000g} \to 1\ \$/kg = 0.1\ ¢/g$

The price is: $30 \times 0.1 = 3$ cent per kilogram.

43) Answer: C

$3P = 6 \to P = 2$ The price of 2007 is one-third of 2012 ($6 \div 3 = 2$)

44) Answer: D

First equation: $4y + 5 = 20x + 9 \to 4y = 20x + 4 \to y = 5x + 1 \to m_1 = 5$

Second question: $5y + 4 = x + 7 \to 5y = x + 3 \to y = \frac{1}{5}x + \frac{3}{5} \to m_2 = \frac{1}{5}$

$m_1 = 5$ and $m_2 = \frac{1}{5}$, they aren't equal slopes or negative reciprocals.

45) Answer: C

Area = $5\frac{1}{3} \times 8\frac{2}{5} = \frac{16}{3} \times \frac{42}{5} = \frac{672}{15} = \frac{672}{15}$ square feet.

One-fourth = $\frac{672}{15} \div 4 = \frac{224}{5} \times \frac{1}{4} = \frac{224}{20} = \frac{112}{10} = 11.2$

46) Answer: A

Two points are $(2,2)$ and $(-4,-1) \to m = \frac{y_2 - y_1}{x_2 - x_1} = \frac{-1-2}{-4-2} = \frac{-3}{-6} = \frac{1}{2}$

47) Answer: B

$V = \left(\frac{base \times height}{2}\right) \times$ height of prism $\to V = \left(\frac{10 \times 15}{2}\right) \times 5 = 375$

48) Answer: C

$f(a) = 4a - 7$ and $f(a) = -19$: $4a - 7 = -19$ so that $4a = -12$ and $a = -3$.

$f(b) = 4b - 7$ and $f(b) = 9$: $4b - 7 = 9$ so that $4b = 16$ and $b = 4$

Finally, f(a + b) = f(−3 + 4) = f(1)

f(1) = 4(1) − 7 = −3.

49) Answer: D

The ratio of $x:y$ is equivalent to x divided by y, or $\frac{x}{y}$.

Equation: $28x = 12y$

Dividing both sides by 28: $x = \frac{12y}{28}$

Dividing both sides by y: $\frac{x}{y} = \frac{12}{28} = \frac{3}{7}$

Then, $x:y$ is $3:7$

50) Answer: B

The sum of exterior angles of any polygon is always equal to 360 degrees.

Exterior angle(degrees) × n = 360; (n, is the number of angles (sides)).

Exterior angle = 180 − interior angle

The exterior angle = 180 − 168 = 12 degrees

Then, $12\,n = 360 \rightarrow n = 360 \div 12 = 30$

The polygon has 30 angles and 30 sides.

THEA Practice Test 5
Answers and Explanations

1) Answer: C

$\frac{-26}{3} = -8.7$ and $\frac{75}{8} = 9.4$, then the even numbers are:

$(-8, -6, -4, -2, 0, 2, 4, 6, 8)$

2) Answer: D

The distance between two points always is positive. Use formula:

AB = |b − a| or |a − b| → |−8 − 7| or |7 − (−8)| = |7 + 8|

3) Answer: B

Sum of the measures of the angles of a triangle is 180, if two angles are 60 then the third one is 60, then the triangle is equilateral triangle, and all side are equal.

4) Answer: B

4 more: +4

Ratio: ÷ ; Ratio of a number to 5: $\frac{x}{5}$

7 less: −7 ; 7 less than the number: $x - 7$

$4 + \frac{x}{5} = x - 7$

5) Answer: A

Rewriting each fraction with common denominator or converting each fraction to decimal and order the decimal from least to greatest.

$\frac{3}{7} = 0.43$ $\frac{5}{9} = 0.56$ $\frac{1}{3} = 0.33$ $\frac{19}{21} = 0.9$ $\frac{11}{18} = 0.61$

6) Answer: B

calculating Elena's total earnings:

35 hours × $9.20 an hour = $322

Next, divide this total by her brother's hourly rate:

$322 ÷ $11.50 = 28 hours

7) Answer: C

number of fiction books: x

number of nonfiction books: $x + 1,200$

Total number of books: $x + (x + 12,000)$

40% of the total number of books are fiction, therefore:

$40\%[x + (x + 1,200)] = x \to 0.4(2x + 1,200) = x$

$0.8x + 480 = x \to 480 = x - 0.8x \to 0.2x = 480$

$\to x = 2,400$ number of fictions

$x + 1,200 = 2,400 + 1,200 = 3,600$, the number of nonfiction books

$2,400 + 3,600 = 6,000$, the total number of books in the library

8) Answer: B

The scale is: 5 cm:1km, (5 cm on the map represents an actual distance of 1 km).

first necessary to rewrite the scale ratio in terms of units squared:

5^2 cm square:1^2 km square, which gives $25\ cm^2 : 1\ km^2$

Then, $\frac{25\ cm^2}{1 km^2} = \frac{80\ cm^2}{x\ km^2}$, (where x is the unknown actual area).

Every proportion you write should maintain consistency in the ratios described (km^2 both occupy the denominator).

Cross-multiply and isolate to solve for the unknown area x:

$x \cdot \frac{25\ cm^2}{1 km^2} = 80\ cm^2 \to x = 80\ cm^2 \cdot \frac{1\ km^2}{25\ cm^2} \to x = 3.2\ km^2$

9) Answer: B

move decimal point in divisor so last digit is in the unit place (0.4 to 4)

move decimal point in dividend same number of places to the right, (12.24 to 122.4)

divide (1224 ÷ 04)=306

insert a decimal point into the answer above the decimal point in the dividend (30.6)

THEA Practice Tests

10) Answer: B

Circumference = $2\pi r = 2 \times \pi \times 8 = 16\pi$

11) Answer: C

Be careful with the conversion factor (per hundred gallons; NOT per gallon).

$23,700 \times \frac{0.95}{100} = 225.15$

$225.15 + 4.20 = 229.35$

12) Answer: A

$\frac{(x^3 - x^2)}{(x^2 - x)} = \frac{x^2(x-1)}{x(x-1)} = x$

13) Answer: C

For odd index we can have negative radicand.

In the even index, negative radicand is undefined.

$\sqrt{-81}$ has a negative number under the even index, so it is non-real.

Negative numbers don't have real square roots, because negative and positive integer squared is either positive or 0.

14) Answer: D

$3f(2a) = 540 \rightarrow$ (divide by 3): $f(2a) = 180$

(substitute 2a) $5(2a)^2 = 180 \rightarrow$ (divide by 5): $(2a)^2 = 36 \rightarrow 4a^2 = 36 \rightarrow$

$a^2 = 9 \rightarrow a = 3$

15) Answer: A

2 hours and 30 minutes is 2.5 hour.

$R \times T = D \rightarrow 110 \times 2.5 = D \rightarrow D = 275$ miles

16) Answer: B

$10x - 7 \leq 6x + 5 \rightarrow$ Add 7: $10x - 7 + 7 \leq 6x + 5 + 7 \rightarrow$

subtract $6x$: $10x - 6x \leq 12 \rightarrow -4x \leq 12 \rightarrow x \geq -3$

17) Answer: B

Use formula to raise a number: $(x^a)^b = x^{ab} \rightarrow 3^4 = (3^2)^2 = 9^2$

THEA Practice Tests

18) Answer: C

Simple interest rate: I = prt (I = interest, p = principal, r = rate, t = time)

$I = 3{,}800 \times 0.045 \times 3 = 513$

19) Answer: D

A straight angle is an angle measured exactly $180°$

$$55° + 30° = 85°$$
$$180° - 85° = 95°$$

20) Answer: C

$$\frac{32x^5 y^7 z^{-2}}{12\, x^2 y^9 z^0} = \frac{32}{12} \times \frac{x^5}{x^2} \times \frac{y^7}{y^9} \times \frac{z^{-2}}{z^0} = \frac{8}{3} \times x^3 \times \frac{1}{y^2} \times \frac{1}{z^2} = \frac{8x^3}{3\, y^2 z^2}$$

21) Answer: B

$m = \frac{y_2 - y_1}{x_2 - x_1} = \frac{2-(-1)}{4-2} = \frac{3}{2}$

$y - y_1 = m(x - x_1) \rightarrow y - (-1) = \frac{3}{2}(x - 2)$

$y + 1 = \frac{3}{2}(x - 2) \rightarrow 2(y+1) = 3(x-2) \rightarrow 2y + 2 = 3x - 6$

$2y - 3x = -6 - 2 \rightarrow 2y - 3x = -8$

22) Answer: C

$\frac{x^2 + 5}{y + 1} = \frac{(-2)^2 + 5}{2 + 1} = \frac{9}{3} = 3$

23) Answer: C

Ethan paid $3.5 for two hours and $0.75 for each of three half-hour period after that.

$3 \times 0.75 = 2.25$

$3.5 + 2.25 = 5.75$

24) Answer: A

Cross multiply and isolate x: $\frac{c-d}{dx} = \frac{2}{3} \rightarrow 3(c-d) = 2dx \rightarrow x = \frac{3(c-d)}{2d}$

$x = \frac{3}{2}\left(\frac{c}{d} - \frac{d}{d}\right) = \frac{3}{2}\left(\frac{c}{d} - 1\right)$

THEA Practice Tests

25) Answer: D

$x^3 - x = x(x^2 - 1) = x(x-1)(x+1)$

26) Answer: B

Factoring: $(x-4)(x-6) = 0 \rightarrow \begin{cases} x - 4 = 0 \rightarrow x = 4 \\ x - 6 = 0 \rightarrow x = 6 \end{cases}$

27) Answer: A

Two points are (0,2) and (6,5)

$m = \frac{y_2 - y_1}{x_2 - x_1} = \frac{5-2}{6-0} = \frac{3}{6} = \frac{1}{2}$

$y - y_1 = m(x - x_1) \rightarrow y - 2 = \frac{1}{2}(x - 0)$

$y - 2 = \frac{1}{2}x \rightarrow y = \frac{1}{2}x + 2$

28) Answer: D

$\begin{cases} 5 \times (-3x + 4y = -5) \\ 4 \times (2x - 5y = 8) \end{cases} \rightarrow \begin{cases} -15x + 20y = -25 \\ 8x - 20y = 32 \end{cases}$ →add two equations:

$-7x = 7 \rightarrow x = -1$

29) Answer: A

Triangle third side rule: length of the one side of a triangle is less than the sum of the lengths of the other two sides and greater than the positive difference of the lengths of the other two sides.

the third side is less than 5+8=13 and greater than 8-5=3

30) Answer: D

Mean = (118 + 134 + 148 + 151 + 159 + 184) ÷ 6 = 894 ÷ 6 = 149

Only 159 can increase the mean.

31) Answer: C

The mode is the value which occurs with the greatest frequency. New York and Ohio are greatest and the same frequency (3 times).

THEA Practice Tests

32) Answer: B

convert the given distance, 6 km, into centimeters, (units on the map)

6 km = 6,000 m = 600,000 cm

divide by the ratio 1:50,000.

$\frac{600,000}{50,000} = 12$ cm

33) Answer: D

Area of square is: $a^2 = (\sqrt{5})^2 = 5$

34) Answer: B

$2x^4y^3 (4x^3y^0)^3 = 2x^4y^3(64x^9) = 128x^{13}y^3$

35) Answer: C

Change percent to decimal: 18% = 0.18

$0.18 \times 250 = 45$

36) Answer: D

Using the formula for mean:

Mean = $\frac{sum\ of\ the\ several\ given\ values}{number\ of\ value\ given}$

$\frac{x+(x-3)+(x-3)}{3} = \frac{3x-6}{3} = \frac{3(x-2)}{3} = x - 2$

37) Answer: C

a value that is inversely proportional to another value:

$a = \frac{k}{2b-5}$ (Where k is a constant of proportionality)

substitute a and b: $15 = \frac{k}{2(7)-5} \rightarrow k = 15(9) = 135 \rightarrow a = \frac{135}{2b-5}$

38) Answer: B

Baseballs and basketballs are spherical. The volume of sphere: $v = \frac{4}{3}\pi r^3$

$36\pi = \frac{4}{3}\pi r^3 \rightarrow 36 = \frac{4}{3}r^3 \rightarrow 108 = 4r^3 \rightarrow r^3 = 27 \rightarrow r = 3$

$d = 2r \rightarrow d = 2 \times 3 = 6$

THEA Practice Tests

39) Answer: A

Use percent formula: Part $= \frac{\text{percent} \times \text{whole}}{100} \Rightarrow 221 = \frac{\text{percent} \times 170}{100}$

$\Rightarrow \frac{221}{1} = \frac{\text{percent} \times 170}{100}$, cross multiply. $22{,}100 = \text{percent} \times 170$,

divide both sides by 170. → percent = 130

40) Answer: D

The sample shows that 3 out of 200 phones will be faulty. Consequently, a proportion can be set up.

$\frac{3}{200} = \frac{P}{12{,}000}$ (P: Faulty Phone) → $P = \frac{3 \times 12{,}000}{200} = 180$

41) Answer: B

We can Write an equation to solve the problem.

Emma Books = Mia books + 5 → Mia book = Emma − 5 = b−5

Emma + Mia = 17

$b + b - 5 = 17 \to 2b - 5 = 17 \to 2b = 17 + 5 \to b = 11$

42) Answer: D

Probability $= \frac{\text{number of desired outcomes}}{\text{number of total outcomes}} = \frac{4}{5+4+7} = \frac{4}{16} = \frac{1}{4} = 0.25$

43) Answer: A

Surface Area of a cylinder $= 2\pi r (r + h)$,

The radius of the cylinder is 6 inches, and its height are 9 inches. π is 3.14. Then:

Surface Area of a cylinder $= 2(3.14)(6)(6 + 9) = 565.2$ inches

44) Answer: D

A line perpendicular to a line with slope m has a slope of $-\frac{1}{m}$.

So, the slope of the line perpendicular to the given line is $-\frac{1}{\frac{c}{d}} = -\frac{d}{c}$.

45) Answer: C

If 38% of the total number of customers is female, then $100\% - 38\% = 62\%$ of the customers are male. Calculate 62% of the total. $0.62 \times 950 = 589$.

THEA Practice Tests

46) Answer: B

First, add the numbers given in the proportion to get a denominator: $8 + 3 = 11$.
Then, the two parts can be represented as $\frac{8}{11}$ and $\frac{3}{11}$.

The smaller share is $\frac{3}{11}$ of 54.5 kg: $\frac{3}{11} \times 54.5\ kg = 14.86$

47) Answer: A

Since this is subtraction of 2 fractions with different denominators, their least common denominator is: $(x - 2)(3x + 1)$

$$\frac{5}{(x-2)} - \frac{6}{(3x+1)} = \frac{5(3x+1)-6(x-2)}{(x-2)(3x+1)} = \frac{15x+5-6x+12}{(x-2)(3x+1)} = \frac{9x+17}{(x-2)(3x+1)}$$

48) Answer: C

State the problem in a mathematical equation:

$2x = 5x - 18 \rightarrow 2x - 5x = -18 \rightarrow -3x = -18 \rightarrow x = \frac{-18}{-3} = 6$

$5x - 18 = 5(6) - 18 = 12$

49) Answer: C

$3h = 2p + 2$:

$p = 2 \rightarrow 3h = 2(2) + 2 = 6 \rightarrow 3h = 6 \rightarrow h = 2$

$p = 5 \rightarrow 3h = 2(5) + 2 = 12 \rightarrow 3h = 12 \rightarrow h = 4$

$p = 11 \rightarrow 3h = 2(11) + 2 = 24 \rightarrow 3h = 24 \rightarrow h = 8$

possible values of h: {2,4,8}

50) Answer: A

There are 6 digits in the repeating decimal (142857), so 1 would be the first, seventh, thirteenth digit and so on.

To find the 391st digit, divide 391 by 6. $391 \div 6 = 65\ R1$

Since the remainder is 1, that means the 391st digit is the same as the 1st digit, which is 1.

THEA Practice Test 6
Answers and Explanations

1) Answer: C

The smallest prime number is 2, and the largest even negative integer is −2.

2+3 (−2) = 2−6 = −4.

2) Answer: B

The difference between his income and his cost is monthly saving.

January: $3,025 − $1,870 = $1,155

February: $3,405 − $1,916 = $1,489

March: $3,280 − $1,962 = $1,318

3) Answer: D

State the problem in a mathematical sentence:

$a + 42 = 230 - 3a$

$a + 3a = 230 - 42$

$4a = 188 \rightarrow a = 47$

4) Answer: A

$4\frac{1}{21} - 5\frac{4}{7} + 2\frac{1}{3} = (4 - 5 + 2)\frac{1}{21} - \frac{12}{21} + \frac{7}{21} = 1\frac{1}{21} - \frac{5}{21} = \frac{22}{21} - \frac{5}{21} = \frac{17}{21}$

5) Answer: C

$3^3 + 3^2 + 3 = 27 + 9 + 3 = 39$

6) Answer: D

$\frac{108}{270} = \frac{108}{27} \times \frac{1}{10} = 4 \times \frac{1}{10} = 0.4$

7) Answer: A

Use the formula for Percent of Change:

$\frac{New\ Value - Old\ Value}{Old\ Value} \times 100\% = \frac{31.50 - 30}{30} \times 100\% = \frac{1.50}{30} \times 100\% = 5\%$

THEA Practice Tests

8) Answer: A

$|2x - 3| = 5 \to \begin{cases} 2x - 3 = 5 \to 2x = 8 \to x = 4 \\ 2x - 3 = -5 \to 2x = -2 \to x = -1 \end{cases}$

9) Answer: C

Multiply equation (2) by 3. Add two equations [(1) +3(2)]:

$\begin{cases} x - 3y = 1 \\ 6x + 3y = 6 \end{cases} \to 7x = 7 \to x = 1$

Substitute $x = 1$ into equation (1): $1 - 3y = 1 \to -3y = 0 \to y = 0$

10) Answer: B

2 hours equals 120 minutes. The train passes 4 stations every 15 minutes: $\frac{4}{15}$

$\frac{4}{15} = \frac{x}{120} \to x = \frac{120 \times 4}{15} = 32$ stations

11) Answer: D

Difference in percent: $8\% - 6\% = 2\%$

$2\% \times 180 = 0.02 \times 180 = 3.60$

12) Answer: C.

$C = \pi d = 2\pi r = 2\pi \times 43 = 86\pi$

13) Answer: C

$V = l \times w \times h = 19 \times 12 \times 9 = 2{,}052 \; in^3$

14) Answer: D

$\sqrt[4]{4^{-8}} = \sqrt[4]{\frac{1}{4^8}} = \frac{\sqrt[4]{1}}{\sqrt[4]{4^8}} = \frac{1}{4^{\left(\frac{8}{4}\right)}} = \frac{1}{4^2} = 4^{-2} = \frac{1}{16}$

15) Answer: A

There are no values of the variable that make the equation true.

16) Answer: C

Volume of cylinder: $V = \pi r^2 h = \pi \times 8^2 \times 12 = 768\pi$

Volume of cone: $V = \frac{1}{3}\pi r^2 h = \frac{1}{3}\pi \times 8^2 \times 6 = 128\pi$

$768\pi + 128\pi = 896\pi$.

THEA Practice Tests

17) Answer: D

$3.21 \times 10^{-4} = 0.000321$

18) Answer: C

$x + 53 + 32 + 45 = 180 \rightarrow x + 130 = 180 \rightarrow x = 50$

19) Answer: B

use the Pythagorean theorem to find the value of unknown side.

$a^2 + b^2 = c^2 \rightarrow 20^2 = a^2 + 12^2 \rightarrow a^2 = 400 - 144 = 256 \rightarrow a = 16$

20) Answer: C

Use percent formula: Part $= \frac{\text{percent} \times \text{whole}}{100} \rightarrow$ Part $= \frac{18 \times 250}{100} = 45$

Last price: $250 + 45 = \$295$

21) Answer: A

Area of trapezoid: $\frac{(a+b)}{2} \times h \rightarrow A = \frac{(12+14)}{2} \times 9 = 117$

Area of triangle: $\frac{b \times h}{2} \rightarrow A = \frac{8 \times 9}{2} = 36$

Area of shaded region: $117 - 36 = 81$

22) Answer: D

$xy - 5x = 32 \rightarrow x(y - 5) = 32$

$8x = 32 \rightarrow x = 4$

23) Answer: C

Multiply both sides by 3: $3 \times \left(\frac{x}{3}\right) = 3 \times (x - 4) \rightarrow x = 3x - 12$

Subtract both side by x and add 12 to both sides: $12 + x - x = 3x + x - 12 +$

$12 \rightarrow 4x = 12 \rightarrow x = 3$

24) Answer: D

$\begin{cases} 3 \times (-5x + 2y = 20) \\ -2 \times (-4x + 3y = 9) \end{cases} \rightarrow \begin{cases} -15x + 6y = 60 \\ 8x - 6y = -18 \end{cases}$ →add two equations:

$-7x = 42 \rightarrow x = -6$

THEA Practice Tests

25) Answer: A

There are 3 parts labeled "F" out of a total of 8 equal parts.

The probability of not spinning at "F" is 5 out of 8.

26) Answer: C

The best estimate of the product of 23 and 2.08 to the nearest whole number.

Since 23 is already a whole number and 2.08 is closer to 2 than 3. Option C is the correct answer.

27) Answer: D

Fuel consumption rate $= \frac{12}{75} = 0.16$ liter per mile

Cost: $1,700 \times 0.16 \times 1.05 = 285.60$

28) Answer: B

Point $1(x_A, y_A) = (-2, -7)$; Point $2(x_B, y_B) = (6, -1)$

Distance between two points $= \sqrt{(x_B - x_A)^2 + (y_B - y_A)^2}$

$\rightarrow d = \sqrt{(6-(-2))^2 + (-1-(-7))^2} = \sqrt{8^2 + 6^2} = \sqrt{64 + 36}$

$\rightarrow d = \sqrt{100} = 10$

29) Answer: A

Ordered values: 122, 122, 165, 186, 212, 318, 334.

The 4th number is median.

30) Answer: D

$\sqrt{3x - 1} = 4 \rightarrow 3x + 1 = 16 \rightarrow 3x = 15 \rightarrow x = 5$

31) Answer: A

Get the equation into slop intercept form:

$y = mx + b$ where m is slope and b is y-intercept.

$x = 2y - 4$; Add 4 to both sides $\rightarrow x + 4 = 2y$; Dividing by 2: $\rightarrow y = \frac{1}{2}x + 2$

Thus, the y-intercept is 2.

THEA Practice Tests

32) Answer: C

$3^{2x} = 81 \rightarrow (3^2)^x = 81 \rightarrow 9^x = 9^2 \rightarrow x = 2$

33) Answer: C

$24: 2 \times 2 \times 2 \times 3$

$30: 2 \times 3 \times 5$

LCM $(30, 24) = 2 \times 2 \times 2 \times 3 \times 5 = 120$

34) Answer: B

Surface area $(SA): 6a^2 \rightarrow 54 = 6a^2 \rightarrow a^2 = 9 \rightarrow a = 3$ cm

Volume $(v): a^3 \rightarrow v = 3^3 = 27$ cm^3

35) Answer: D

Write the equation into standard form: $x^2 + x = 30 \rightarrow x^2 + x - 30 = 0$

Factor this expression: $(x-5)(x+6) = 0 \rightarrow x = 5$ or, $x = -6 \rightarrow x^2 = \begin{cases} 25 \\ 36 \end{cases}$

36) Answer: D

The minimum amount of water: $5 \times \frac{3}{5} = 3; \; 3 \div 2 = 1.5$

The maximum amount of water: $5 \times \frac{3}{5} = 3$

Amount of water in all pitchers: $1.5 < w < 3$

37) Answer: C

The range is: highest value – lowest value.

R $= 45 - 28 = 17$

38) Answer: A

d $= 2r \rightarrow r = \frac{d}{2} = \frac{14}{2} = 7$

Area of circle (A): $\pi r^2 = \pi \times 7^2 = 153.9$

39) Answer: B

Factor the expression: $\frac{2x^2-7x-4}{4(x^2-\frac{1}{4})} = \frac{(x-4)(2x+1)}{4x^2-1} = \frac{(x-4)(2x+1)}{(2x-1)(2x+1)} = \frac{x-4}{2x-1}$

WWW.MathNotion.com

THEA Practice Tests

40) Answer: D

$(\frac{1}{3})^{-3} = \frac{1}{3^{-3}} = 3^3 = 27$

41) Answer: B

$\frac{(5x^{-2}y^3)^2}{125y^{-2}z^{-1}} = \frac{25x^{-4}y^6}{125y^{-2}z^{-1}} = \frac{y^8 z}{5x^4}$

42) Answer: D

Rate: $\frac{\$1}{1Kg} = \frac{100¢}{1,000g} \rightarrow 1\,\$/kg = 0.1\,¢/g$

The price is: $10 \times 0.1 = 1$ cent per kilogram.

43) Answer: C

$2P = 2 \rightarrow P = 1$ The price of 2007 is half of 2012 ($2 \div 2 = 1$)

44) Answer: D

First equation: $2y - 1 = 4x + 5 \rightarrow 2y = 4x + 6 \rightarrow y = 2x + 3 \rightarrow m_1 = 2$

Second question: $4y - 1 = 2x + 5 \rightarrow 4y = 2x + 6 \rightarrow y = \frac{1}{2}x + \frac{3}{2} \rightarrow m_2 = \frac{1}{2}$

$m_1 = 2$ and $m_2 = \frac{1}{2}$, they aren't equal slopes or negative reciprocals.

45) Answer: C

Area = $9\frac{1}{2} \times 7\frac{1}{5} = \frac{19}{2} \times \frac{36}{5} = \frac{684}{10} = \frac{342}{5}$ square feet.

One-fourth = $\frac{342}{5} \div 4 = \frac{342}{5} \times \frac{1}{4} = \frac{342}{20} = \frac{171}{10} = 17.10$

46) Answer: A

Two points are $(6,6)$ and $(-10,-2) \rightarrow m = \frac{y_2 - y_1}{x_2 - x_1} = \frac{-2-6}{-10-6} = \frac{-8}{-16} = \frac{1}{2}$

47) Answer: B

$V = \left(\frac{base \times height}{2}\right) \times$ height of prism $\rightarrow V = \left(\frac{9 \times 15}{2}\right) \times 4 = 270$

48) Answer: C

$f(a) = 5a - 2$ and $f(a) = -12$: $5a - 2 = -12$ so that $5a = -10$ and $a = -2$.

$f(b) = 5b - 2$ and $f(b) = 13$: $5b - 2 = 13$ so that $5b = 15$ and $b = 3$

THEA Practice Tests

Finally, f(a + b) = f(−2 + 3) = f(1)

f(1) = 5(1) − 2 = 3.

49) Answer: D

The ratio of $x:y$ is equivalent to x divided by y, or $\frac{x}{y}$.

Equation: $8x = 6y$

Dividing both sides by 8: $x = \frac{6y}{8}$

Dividing both sides by y: $\frac{x}{y} = \frac{6}{8} = \frac{3}{4}$

Then, $x:y$ is $3:4$

50) Answer: B

The formula for measurement of each angle of a regular polygon:

$x = \frac{180(n-2)}{n}$, (x is the measurement of the interior angle; n is the number of sides)

Substituting the given information:

$\frac{180(n-2)}{n} = 144 \rightarrow 180n - 360 = 144n \rightarrow 180n - 144n = 360$

$\rightarrow 36n = 360 \rightarrow n = 10$

The polygon has 10 angles and 10 sides.

"END"

www.ingramcontent.com/pod-product-compliance
Lightning Source LLC
Chambersburg PA
CBHW081323040426
42453CB00013B/2285